STELLAR EVOLUTION

STELLAR EVOLUTION

Amos Harpaz

Scientific Editor
Oded Regev

 CRC Press
Taylor & Francis Group
Boca Raton London New York

CRC Press is an imprint of the
Taylor & Francis Group, an **informa** business

AN A K PETERS BOOK

First published 1994 by A K Peters, Ltd.

Published 2018 by CRC Press
Taylor & Francis Group
6000 Broken Sound Parkway NW, Suite 300
Boca Raton, FL 33487-2742

ISBN-13: 978-1-56881-012-6 (hbk)

Library of Congress Cataloging-in-Publication Data

Harpaz, Amos.
  Stellar evolution / Amos Harpaz.
    p. cm.
  Includes bibliographical references and index.
  ISBN 1-56881-012-1
  1. Stars — Evolution. 2. Nuclear astrophysics. I. Title.
QB806.H37   1993
523.8 — dc20   93-38857
                                           CIP

About the cover: A processed image of the circumstellar ring of the supernova SN 1987A. The image was prepared from an observation by the Hubble Space Telescope in August 1990. The luminous ring is created by the interaction of the radiation emitted by the central star (the supernova), with the matter residing around the supernove. This matter was ejected from the central star about 40,000 years ago, and the distance of this matter from the central star is about one light year. (Panagia et al., 1991, *Astrophys. J. Let.*, **380**, L23.) Details are given in Chapter 10, Section 10.4.

Cover image courtesy of the Space Telescope Science Institute

Contents

List of Figures

Physical and
Astronomical Constants

Gravitational constant	G	6.671×10^{-8} dyne cm^2 gm^{-2}
Speed of light	c	2.998×10^{10} cm sec^{-1}
Boltzmann constant	k	1.381×10^{-16} erg degree^{-1}
Stefan–Boltzmann constant	σ	5.67×10^{-5} erg cm^{-2} sec^{-1} degree^{-4}
Planck's constant	h	6.6262×10^{-27} erg sec
Solar mass	$M_\odot$	1.991×10^{33} gm
Solar luminosity	$L_\odot$	3.86×10^{33} erg sec^{-1}
Solar radius	$R_\odot$	6.96×10^{10} cm
Earth's mass	$M_\oplus$	5.98×10^{27} gm
Light year	LY	9.4605×10^{17} cm

Introduction

Stellar Evolution presents the theory of the evolution of stars on an undergraduate college level, and can be used as a textbook in the subject.

Historically, interest in celestial objects has followed two parallel courses. In one, people gazed at and were impressed by regular celestial events, and the occurrence now and then of spectacular events. In the other, people attempted to understand the overall picture of the stellar world and the laws governing its evolution. They tried within this framework to establish a holistic and internally consistent theory that might explain all the observed phenomena. The kind of thinking involved in the first course yielded many useful and efficient tools. Determining and monitoring the calendar and using celestial objects for purposes of navigation are just a couple of the important applications. The thinking involved in the second course is part of the common effort of humanity to understand the world in which we live, and to construct a general world picture of reality. The fact that celestial objects are in practical terms "beyond the reach of human hands" establishes a special place for them in the picture of nature. From the time of the ancient civilizations until today, the theory of celestial objects has played a principal role in constructing a general world view.

Obviously, these two courses of thought are interrelated; any advancement in one of them has had an important impact on the other. Broadly speaking, the first course of thought gave birth to astronomy whereas the second gave rise to astrophysics and cosmology.

The present book deals with astrophysics or the theory of stellar structure and stellar evolution. The first half of the book introduces this theory, along with details of the physical laws which govern stellar evolution. The theory is then used to present the evolutionary track of stars. In the second half of the book, the theory is used to understand a number of interesting and important phenomena involved in stellar evolution. Special emphasis is given to examples taken from the most recent astrophysical and astronomical studies, with the intention of acquainting the reader with the up-to-date research. With the rapid development and improvement of observational instruments, and our present ability to carry out observations above the

1

atmosphere, numerous discoveries have been made and many more can be expected in the near future. Exposure to the work carried out today will enable the reader to follow and understand the new findings that await us tomorrow.

These examples will also demonstrate how astrophysicists work, what information is available to them, and the way in which they use this information in their effort to construct a complete picture. Usually the information available to astrophysicists is partial and fragmentary, because our observations of astronomical objects and astrophysical events are passive. Scientists cannot initiate experiments in astrophysics and can only observe and follow the flow of events which nature displays. Moreover they cannot "illuminate" a stellar object or an astrophysical event in which they are interested, so as to observe it with greater precision. In order for an event to be observed, it must emit a characteristic radiation in the direction, and with a power, which makes it detectable by instruments. Indeed, recent developments in instrumentation (which can now cover the whole range of wavelengths with very high sensitivity), as well as our ability to place telescopes of various kinds above the atmosphere (thereby avoiding most of the disturbing problems we encounter on Earth), have widened by a few orders of magnitude the quantity and range of characteristics of information available to astrophysicists. They have discovered many new phenomena and accumulated more details about known phenomena as a result. Thus, our understanding of the stellar world is better established and more comprehensive than ever before.

Nevertheless, large gaps in our information still exist. An important part of an astrophysicist's work is to bridge these gaps by a careful analysis of available information. By using creative imagination, missing links can be deduced. Some of the examples presented in this book illustrate how a complete picture is constructed from the separate pieces of the puzzle. The tests for a satisfactory theory are its completeness, its consistency, and the agreement of its consequences with the observed data.

The basic theory is presented in Chapters 2 to 4. The methods of computation which are used in stellar evolution are explained in Chapters 4 and 5. Readers who are familiar with these topics — or, alternatively, who are not interested in them — may skip these and begin reading from the second half of Chapter 5. Starting at this point will still furnish the reader with a general view of the subject, as well as an understanding of the current work and achievements in astrophysics. The narrative part of the book can be understood without going into all the details of the equations and mathematical tools which are laid out in the book. However for a better understanding of the subject, a systematic study of the material is recommended.

Chapter 1 defines the general framework of astrophysics, and presents some important aspects of the field. The equations for calculating the structure of a star are laid out in Chapter 2: the equation of hydrostatic equilibrium, the equation of heat transport, the energy equation, and the equation of mass continuity. These equations form the base from which the structure of a star is derived. The equations of the physics of the stellar interior are complemented by the formulae of gas characteristics presented in Chapter 3: the equation of state for different situations of matter; the opacity of matter as a function of the composition and the thermodynamic state; and the equations for energy production through nuclear reactions.

Using the physics described in the preceding chapters, and presenting the basic numerical methods employed in calculating stellar structure, models of stars are constructed in Chapter 4. Here the reader is introduced to models of a main sequence star, a red giant, and a white dwarf. Chapter 5 defines the principles needed to calculate stellar evolution and presents the methods by which these calculations are accomplished. Chapter 6 sketches the evolutionary track of a star, following the different stages in this evolution and a star's path through them. The observational aspects of the different stages are discussed as well.

Since we know that most stars are part of binary systems, we present in Chapter 7 the basic physics of such systems, together with an example that demonstrates how the observed properties of the stars participating in a binary system can be used to expand our knowledge about stars.

The chapters that follow Chapter 7 are each devoted to a special topic in the study of stellar evolution. In Chapter 8 we treat the topic of star formation. By examining examples of stars existing in the first stages of their life and phenomena connected with this evolutionary phase, such as T Tauri stars and Herbig-Haro objects, we arrive at the general theory that explains how a star is formed from a collapsing cloud of gas. The chapter closes with a description of the formation of the solar system. In Chapter 9 we deal with models of rotating stars, and try to estimate the role of rotation in stellar evolution by investigating how much a rotating model differs from a static one. This chapter concludes with a study of solar rotation.

Chapter 10 deals with the phenomenon of supernova explosions. This is a very extreme phenomenon which changes the star entirely. However since this type of explosion is a very prominent event, it has been extensively studied. A reasonably coherent picture of the entire process leading to the explosion is now available as a result. A description of the recent supernova explosion of SN 1987A closes the chapter. In Chapter 11 we return to the topic of binary systems, and deal specifically with close binary systems. Owing to the small separation between the components in such systems, the mutual interaction between the companions is significant. Mass transfer may take place at a high rate, resulting in interesting changes in the evolutionary course of the stars and in the emission of specific types of radiation created by the accreted mass. One branch of such an activity leads to a nova outburst, which is described in some detail.

In Chapter 12 we deal with special topics, such as: the problem of the solar neutrinos, neutrino cooling, neutron stars, pulsars, SS 433, and Chaos in stellar variability. The book ends with the description in Chapter 13 of the Galaxy, which is regarded as the stage in which the evolution of known stars takes place.

Some preliminary notes are needed to explain the technical concepts used in this book. When discussing nuclear reactions, say of hydrogen, we use the word "burning" of hydrogen. In its ordinary usage the same word is employed for the reaction of an element with oxygen, as in the burning of wood. However in the jargon of astrophysicists, this word is used for the fusion of light elements to heavier ones, whereby nuclear energy is released. We shall be using the word "burning" in this sense.

Astronomical objects are enormous both in their dimensions and in their masses. In order to avoid using huge numbers, it is customary to employ the solar properties as the units for stellar masses, radii, and luminosities. The sign we use for the Sun is the one used by medieval alchemists for the Sun and for gold, $\odot$. Thus, $M_\odot$, $R_\odot$, $L_\odot$, are the solar mass, the solar radius, and solar luminosity respectively. They will be used to represent the masses, radii, and luminosities of stars in this book. In the same way we designate by $\oplus$ quantities related to the Earth, like $M_\oplus$ which represents the Earth's mass. For a unit of distance we use a light year (LY), which is the distance covered by light during a single year: $1 \text{ LY} \simeq 9.46 \times 10^{12}$ km.

Some of the differential equations used in astrophysics are partial differential equations. We are not going to solve such equations in this book. However for the sake of completeness, the equations are written as partial differential equations, where $\frac{\partial}{\partial x}$ denotes a partial derivative with respect to x; and $\frac{d}{dx}$ is the ordinary derivative with respect to the same variable.

It is with pleasure that I acknowledge the help I received from some of my colleagues. The section on Chaos in Stellar Variability was written by Professor O. Regev. Dr. G. Berger read the section on the equation of state and contributed valuable comments. Dr. D. Kovetz-Prialnik read Chapter 11, and her comments proved most helpful as well. Dr. N. Soker read through the entire manuscript, and helped to improve it with his observations. I would like to thank Mr. Ted Gorelick for textual and stylistic editing of the manuscript, and Mr. Hillel Pesach for drawing the artistic view of SS 433.

While writing this book I was a guest researcher at the Department of Physics and the Space Research Institute at the Technion, in Haifa, whose people I wish to thank for their hospitality during this period.

Chapter 1

Astronomy and Astrophysics

A star is by definition a celestial object, bound by self-gravity, in which thermal energy is produced from a nonthermal source (nuclear reactions) and is radiated outward from the object's surface to its surroundings. We shall extend this definition somewhat so as to include objects such as white dwarfs and neutron stars which produced nuclear energy in their past, as well as objects that are in the process of becoming stars (protostars). According to this definition, planets, meteorites, comets, and other similar celestial objects are not stars.

As a science, astronomy has a long history. It has dealt with the nomenclature of stars and with observations of their luminosities, their brightness and colours, and their grouping and motions. It has systematically monitored the changes of their observed properties. Astronomers have carried out regular observations of celestial objects from the time of ancient civilizations. They have observed, mapped, and named single stars and constellations of stars, and have followed and registered with a high degree of precision the "apparent" movements of the planets, the Sun, and the Moon.

Astrophysics, on the other hand, is a very young science. Several stages were needed in the evolution of physics before people could start thinking of stars as physical objects. An important step in this direction was taken when people realized the enormity of the distances to the stars; they concluded that the stars must be very luminous objects on the order of magnitude of the Sun. This stage in our understanding occured in the eighteenth century, when Bessel calculated the distance to the star 61 Cygni by using its measured parallax. His calculations showed that the distance to the star is a few light years, and the stellar luminosity calculated using

7

this distance was about that of the Sun. The study of thermodynamics and the behaviour of matter in different thermodynamic situations led to a picture of a star as being a very hot gaseous sphere, in which gravity and thermodynamic pressure balance each other in a dynamic equilibrium. Astrophysics became an independent science when the researchers began to understand the physical processes that take place inside stars.

1.1 The Energy Source of Stars

The first question scientists tried to answer was: what is the source of the huge amount of energy radiated by a star? The immediate answer that came to mind was that the source was the gravitational energy released during the formation of the star from a dilute gas cloud. We can calculate this energy by noting that the energy of a mass element dm, when added to a mass m from an infinite distance, is $dE = -Gm\,dm/r$, where G is the gravitational constant and r is the radius of the mass m. Assuming that the star is built by adding such mass elements, starting from zero, we can find the energy by integrating over dE. If for the sake of convenience we assume that the density of the mass in the star ρ is constant and uniform, we can substitute for m and dm in dE: $m = \frac{4\pi}{3}r^3\rho$, $dm = 4\pi r^2 \rho dr$. We integrate:

$$E_G = \int_0^E dE = -\frac{G(4\pi\rho)^2}{3}\int_0^R r^4 dr = -G\frac{3}{5}\left(\frac{4\pi}{3}\rho\right)^2 R^5 = -\frac{3}{5}\frac{GM^2}{R}.$$

$$(1.1)$$

A more realistic calculation, which takes into account that the density in a star is not uniform but greater at the centre, yields a higher value for E_G. Dropping the coefficient 3/5, and substituting in this expression the values of the radius and the mass of the Sun, we obtain for the approximate energy released during the contraction:

$$E_G = -\frac{GM^2}{R} = -3.8 \times 10^{48} \text{ erg}. \tag{1.2}$$

The rate of energy radiation by a star like the Sun is $L_\odot = 3.9 \times 10^{33}$ erg sec^{-1}. Dividing E_G by this radiation rate we find the period for which this energy will radiate at the present solar intensity:

$$t = \frac{E_G}{L_\odot} \sim 10^{15} \text{ sec} = 31.7 \text{ million yr}. \tag{1.3}$$

Thus if all the gravitational energy released during the formation of the star is emitted as radiation at the present rate of solar radiation, it may last for about 30 million years. This is how long it takes for a star like our Sun to cool down by radiation. This period is called the thermal time scale or *Kelvin-Helmholtz time*. This is indeed a very long span of time, but fossils of primitive algae found in ancient rocks that are several billion years old show that thermal conditions on the Earth did not change greatly during this period. Since the main source for energy on Earth is the Sun, we conclude that the Sun has existed in generally the same thermodynamic state for a period of several billions of years. We, therefore, have to look for another energy source which could supply much more energy than that furnished by the gravitational contraction of a star. This consideration led scientists to conclude that the energy radiated from the Sun and the stars is supplied by a nonthermal source — namely, by nuclear reactions.

The energy of nuclear reactions is the energy released during the transmutations of light elements to heavier ones. The amount of energy that can be supplied is proportional to the mass available for these transmutations. This can be expressed as an equation which relates the luminosity of a star, L, to the transmutation rate by nuclear reactions, $d(\beta M)/dt$:

$$L \propto -\frac{d(\beta M)}{dt} \qquad (1.4)$$

where β is a number which shows what fraction of the stellar mass is available for the nuclear reactions in the stellar interior. The negative sign in expression 1.4 shows that the radiation is produced with a decrease in the stellar mass available for nuclear reactions. We use a proportion sign (instead of an equal sign) in expression 1.4 because we must multiply the term on the right-hand side by c^2 and by a constant which shows what fraction of the matter involved in the nuclear reactions is converted to energy. In hydrogen burning this fraction is about 0.7 percent. In nuclear reactions of heavier elements the fraction is much smaller.

The mass of a star can be determined only for a star involved in a dynamic interaction with another object, such as another star in a binary system. This topic will be dealt with in detail in Chapter 7. For the moment, we need only mention that the masses of a few hundred stars are determined by using the properties of binary systems. For such stars, whose distance from us can also be determined by using their parallax, we can calculate their luminosity with considerable precision and obtain a mass-luminosity relation. Data about stellar masses and stellar luminosities of about hundred stars presented by Popper[1] are displayed in fig. 1.1.

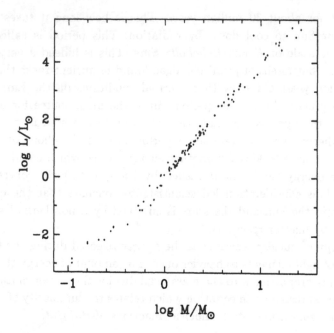

Figure 1.1. Mass-luminosity relation. Note that the scale on both axes is logarithmic.

From this figure we find that the luminosity is proportional to a power of 3.5 of the stellar mass. This can be formulated as:

$$L \propto M^{7/2}. \tag{1.5}$$

The mass available for nuclear transmutation in a given star is limited. Comparing expressions 1.4 and 1.5, we observe that the "intensity of life" of high mass stars is much greater than that of low mass stars. The result is that the lifetime of high mass stars is much shorter than that of low mass stars.

1.2 Hertzsprung-Russell's Diagram

We can treat the temperature of a radiating surface as the *effective temperature* for the radiation, T_e. The radiative flux H (energy radiated per unit time per unit surface) is given by $H = \sigma T_e^4$ erg cm^{-2} sec^{-1}, where σ is the Stefan-Boltzmann constant. Multiplying H of a given star by the stellar surface area, we obtain the total luminosity of the star. The surface area of a spherical object with a radius R is $4\pi R^2$. The luminosity of a star is given by:

$$L = 4\pi R^2 \sigma T_e^4 \quad \text{erg} \quad \text{sec}^{-1}. \tag{1.6}$$

A very useful tool in studying stellar properties is the *Hertzsprung-Russell (H-R) diagram.* In 1912 Hertzsprung and Russell showed that when the information on stars is arranged in a diagram whose axes are the logarithms of their luminosities and their effective temperatures T_e, most of the stars will fall in a virtually straight line. This line is called the *main sequence* (MS). Main sequence stars are found to be those which burn hydrogen at their centres. The vertical axis in the H-R diagram is log L and the horizontal axis is log T_e, decreasing from left to right. A plot of the H-R diagram of the stars displayed in fig. 1.1 is shown in fig. 1.2.

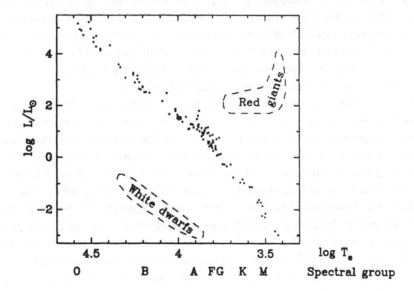

Figure 1.2. H-R diagram. The horizontal axis is scaled here by *log T_e*, while in the original Hertzsprung and Russell diagram it was scaled by the spectral classes.

Details about this diagram can be found in an article by H.N. Russell.[2] The location of a star in the diagram furnishes important information. In the original diagram the horizontal axis was actually arranged according to the spectral classes of the stars. These classes are marked by the letters

O, B, A, F, G, K, M. Moving in this sequence from left to right, we proceed from the blue edge of the spectrum to the red. The connection between the spectral class and the effective temperature of a star became clear later. In fig 1.2 we present both the logarithms of the effective temperature and the spectral class along the horizontal axis. Later, we refer to stars by the names of their spectral classes, remembering that the stars in classes O, B, and A are the hot stars, while those in classes K and M are the cool stars. Subsidiary divisions of a class are denoted by adding decimal numbers from 1 to 10 to each letter. Thus the Sun is marked as a G2 star, meaning that it is in the second subsidiary division of spectral class G.

When the luminosity and effective temperature of a star are known from its location in the diagram, its radius is also known. A star located in the upper right-hand corner of the diagram has low effective temperature (it is red); however its luminosity is high which means that its radius is very large. This is a red giant star. A star located at the lower left-hand corner of the diagram has high effective temperature (it has a hot surface), but because its luminosity is low it has a very small radius. This is a white dwarf star. When we come to discuss the evolutionary tracks of stars, we shall relate them to their locations in the H-R diagram.

The information needed to study the physics of stars is gathered by observation. Several types of observations yield different kinds of information which complement one another in constructing the overall picture of stellar physics. First, we have information about the motion of stars and changes in their physical dimensions. Observation of the orbital motion of binary stars in their rotation furnishes information on the dimensions of the binary system, the velocities of its components, and the separation between the companions. By keeping track of the linear motions of stars and comparing them to those of neighbouring objects, we can determine if stars belong to the same system. We also learn if they are dynamically interconnected, as in a star cluster, or are only accidentally located on the same line of sight. The expansion of stars during nova explosions can be followed by consecutive photographs taken at appropriate time intervals.

Another important type of information about stars is derived from the study of stellar spectra. The chemical composition can be deduced from the characteristic spectral lines of the elements. The relative strength of the lines show on the one hand the relative abundance of the elements and on the other hand the thermodynamic conditions in which these spectral lines developed. The Doppler effect found in the spectral lines indicates the velocity (relative to ourselves) of the source from which these lines formed.

In classifying spectral lines, we distinguish between emission lines and absorption lines. *Emission lines* form when the electrons in the radiating atom drop from high to lower energy levels. The spectral lines correspond to the energy difference between the initial and final energy levels. By observing the emitting matter directly, these emission lines are discerned. A beam of white light includes all the spectral colors. When such a beam passes through a gaseous matter, the atoms absorb the photons from the beam. The absorbed photons are those whose energies are exactly equal to the differences between the energy levels in the absorbing atoms. The spectral lines which correspond to these energy differences are subtracted from the continuous spectrum of the original beam. They appear as dark lines on the continuous spectrum and are called *absorption lines.*

Usually the energy radiated by a star is created deep in the stellar interior. It proceeds through millions of steps of absorption and re-radiation until it finally radiates from the stellar surface. This radiation has a continuous spectrum and, on passing through cool gas which surrounds the star, absorption lines form in the radiation. Thus most spectral lines observed in stellar light are absorption lines. When the matter surrounding a star is sufficiently hot to cause transitions from one atomic energy level to another, we see the radiating source directly and observe emission lines in the spectrum. The presence of emission lines in a stellar spectrum tells us that a hot radiating matter (such as a hot disk) is present around the star.

1.3 Distances to Stars

An important piece of information is the distance to an observed object. However, because of the huge expanses between astronomical objects, we have no direct way of measuring the distance to a star. Sophisticated methods are therefore used to build a cosmic distance ladder in which each step is determined by the preceding one.

The first step in measuring the distance to stars is performed by using their observed parallax. The parallax of a star is its apparent motion relative to the distant background as caused by the motion of the observer. The change in an observed angle to a star due to the observer's motion along some baseline enables us to use a trigonometric method to determine the distance to the star. The longest baseline we can use is formed by our motion around the Sun in the course of half a year. The radius of the Earth's orbit around the Sun is about 150 million kilometres. The minimal angle difference that can be measured with sufficient precision from the earth surface through the atmospheric cover is 0.05″ (the notation ″ denotes arcsecond $= \frac{1}{3600}$ of a degree).

Using the trigonometric relation, and taking as a baseline the radius of the Earth's orbit, a parallax of 0.05″ determines a distance of about 60 light years (LY). Within the volume around us, as defined by this distance, about 700 stars are clearly observed and their distance determined. From the distance and observed luminosity of these stars we can determine their intrinsic luminosity. This sample of stars is sufficiently large for us to deduce some of the statistical properties of stars. For those stars involved in binary systems, the mass can be inferred from the properties of the binary system, and we are in turn able to deduce the mass-luminosity relation.

The next step in the ladder is reached by using the period-luminosity relation observed in a certain type of variable stars called Cepheids. These are bright stars that pulsate with constant periods, their pulsation periods ranging from one day to several tens of days. In 1912 Henrietta Leavitt[3] investigated the relation between the luminosity of such stars and their pulsation period in a group of 25 stars, all of them located in the Small Magellanic Cloud (SMC). The SMC is a small galaxy close to our own Galaxy. Because of the great distance to SMC (150 thousand LY), Leavitt was unable to determine their luminosities. However, due to this same great distance, and relatively small dimensions of the SMC, Leavitt could assume that all the stars in the group undergo the same decrease in their luminosities due to the distance, and that the ratio between their observed luminosities is the same as the ratio between their intrinsic luminosities. She found that there is a clear relation between stellar luminosity and the star's pulsation period.

Figure 1.3 is drawn by using the data given by Leavitt[3] for the 25 Cepheids in SMC. The straight dashed line in the figure was drawn by Leavitt, from which she arrived at the period-luminosity relation. The longer the period of pulsation, the greater is the luminosity. This relation furnishes a scale because measurements of period length do not depend on distance. For relatively close Cepheids, the intrinsic luminosities were determined from their observed luminosities and were used to calibrate the scale. Measurements of periods for distant Cepheids were used together with their luminosities (inferred from their pulsation period), so as to determine distances to the groups of stars containing these Cepheids. Cepheids exist in other galaxies, and their luminosity-period relation is used to determine the distance to these galaxies. In very distant galaxies, Cepheids are unobservable, requiring the use of another step in the cosmic distance ladder.

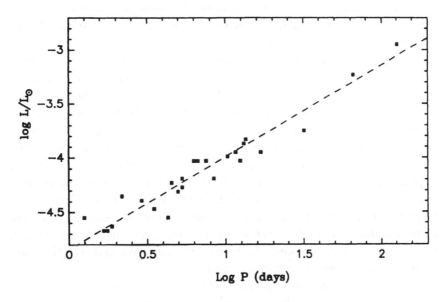

Figure 1.3. Period-luminosity relation of Cepheids.

A detailed treatment of the cosmic distance ladder and its extension to include further steps in space can be found in Weinberg's book.[4]

1.4 Stellar Population

The solar system is located in a large group of stars called the *Galaxy* or the *Milky Way*. Other similar groups of stars are also called galaxies, but ours is *the* Galaxy. The total mass of the Galaxy is about 10^{11} times the solar mass. Assuming that the Sun is a characteristic star, we estimate that there are about 10^{11} stars in the Galaxy. The Galaxy contains a disk where most of the Galaxy's matter is present in the form of stars, gas, and dust. Toward the centre there is a condensation of matter. Gas and dust are the raw materials of star formation. Above and below the disk, there is a spherical distribution of a very dilute stellar population, most of it grouped in *globular clusters*. These are dense groups of tens to hundreds of thousands of stars, dynamically interconnected by mutual gravitational attraction. Part of the stars in the disk are also grouped in clusters called *Galactic clusters*. These are however open, loosely connected clusters, containing fewer stars than would be in a typical globular cluster.

The entire Galaxy rotates, the Galactic disk being the rotational plane, with a period of about 250 million years. The velocity of the Sun due to this rotation is about 200 km sec^{-1}.

Several criteria can be used to distinguish the different types of stellar populations: the kinematic behaviour of stars; the location of stars relative to the Galactic plane and within stellar groups; the abundance of metals in the composition of stars; and the brightness of the brightest stars in a system. By the kinematic behaviour of a star, we mean the star's peculiar velocity. Each star participates as a part of the Galaxy in the ordered motion of Galactic rotation. Superimposed on the ordered motion, which appears as the average motion of the stellar vicinity, there is a peculiar velocity which is the star's velocity relative to its vicinity. When an astrophysicist speaks about metals, he means all the elements heavier than hydrogen and helium. The abundance of metals is calculated as a fraction by mass of the heavy elements from the total mass. We find that there are correlations between these different properties.

First we distinguish the population of the globular clusters and the stars similar to them. These clusters have high peculiar velocities relative to the background population and possess a significant vertical component in their velocity relative to the disk. Many of them are located outside the Galactic plane and have low metal content (about 10^{-3}). The brightest stars in these systems are red dwarfs which are low mass stars. This type of population is called *population II*.

On the other hand we distinguish the population of the Galactic clusters and stars similar to them. They are called *population I*, and are mainly found in the Galactic plane. The peculiar velocities of the stars of this population are low, approximately 10 to 40 km sec^{-1}, and their metal abundance is high — up to four percent. The brightest stars among them are blue and bright, making them high mass stars. These blue, bright stars are usually found close to or inside of the clouds of gas and dust where stellar birth takes place. We believe them to be newly born stars formed in the gas and dust clouds. Such stars then recede from their birthplace. We can estimate the age of these stars from the observation of their recession velocities and present distances from their assumed origin. It has been found that such stars are young, being only about few hundred thousands of years old.

We can generally conclude that extreme population II stars are the oldest stars in the Galaxy, whereas extreme population I stars are the youngest. The composition difference is explained by the assumption that the primordial composition of the matter in the Galaxy consisted mainly of hydrogen and helium and included a very low abundance of metals, and first generation stars had the composition of the primordial matter. During stellar life, part of the light elements is converted to metals. When the stellar matter (or part of it) is returned to the interstellar medium, this medium is enriched by metals. Later generation stars therefore have a higher abundance of metals.

The kinematic behaviour of the different populations is explained by the assumption that the Galaxy formed in a spherical configuration. Dissipation of the gas and dust contained in the Galaxy convert kinetic energy to heat. Owing to this loss of kinetic energy, and due to the rotation of the Galaxy, the configuration takes the form of a disk and the velocities of the constituents decrease. First generation stars, formed while the spherical configuration of the Galaxy still existed, had high velocities and a spherical distribution around the Galactic centre. Stars formed later, after the Galaxy acquired the disk configuration and a part of the kinetic energy converted to heat, had lower velocities relative to the background. These stars were located mainly in the Galactic disk.

From the expressions 1.4 and 1.5, we understand that the intensity of the life of high mass stars is much higher than that of low mass stars. Thus only low mass stars survive today from the first generation of stars, as red dwarfs. High mass stars, appearing as blue bright giants, can only be later generation stars, which are disk-population stars (population I) with high abundance of metallicity. Between the two extremes — extreme population I and extreme population II — we have coverage of the entire range. The population division is summarized in tab. 1.1:

Table 1.1. Stellar population.

Population	Peculiar velocities (km sec^{-1})	Typical members	Metallicity abundance
Young pop. I	10	Blue supergiants	0.04
Intermed. pop. I	20	$\cdots$	0.03
Old pop. I	30	The Sun	0.02
Intermed. pop. II	60	$\cdots$	0.01
Old pop II	> 100	Red dwarfs	0.001

References

1. Popper D.M., 1980, *Ann. Rev. Astron. Astrophys.*, **18**, 115.
2. Russell H.N., 1960, in *A Source Book of Astronomy*, ed. H. Shapley, Harvard University Press, Cambridge, MA.
3. Leavitt H., 1960, in *A Source Book of Astronomy*, ed. H. Shapley, Harvard University Press, Cambridge, MA.
4. Weinberg S., 1972, *Gravitation and Cosmology*, John Wiley & Sons, New York.

Chapter 2

The Equations for Stellar Structure

The evolution of a star from its birth to its death may be schematically described as follows. A diffused cloud of matter starts to contract due to the gravitational force acting between its constituents. If no counterforce develops to stop the contraction, the matter will continue to contract until it collapses to the centre of gravity. But with the contraction, the density and the temperature of the matter increase. The increased density and temperature create pressure which counters the gravitational force and decelerates the contraction. When the force created by the pressure gradient in the star equals the gravitational force, the contraction stops and a hydrostatic equilibrium forms. The density and the temperature gradients create the pressure gradient. Thus during most of stellar life the pressure, density, and temperature are highest at the stellar centre.

If the star were an isolated object which did not lose energy, this balance of forces might remain stable. But the temperature difference between the stellar surface and the space surrounding it causes an energy flow in the form of electromagnetic radiation from the star outwards. This process cools the star. The temperature gradient causes a heat flow from the interior to the surface. As a result of this cooling the pressure decreases, the balance with the gravitational force is disturbed, and the star continues to contract. This sequence of events once again raises the temperature. Increased temperature with increased density again create the pressure gradient which halts the contraction. In the contracted star, the distance of each mass element from the centre of gravity is shorter, and the gravitational force acting on these mass elements is greater. The pressure gradient balancing this force should therefore be steeper. The density and temperature creating the pressure gradient are higher than before the contraction.

19

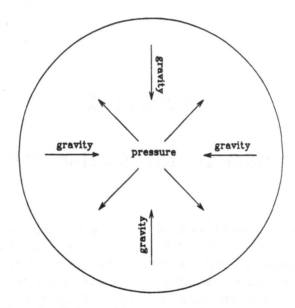

Figure 2.1. Balance of forces in a hydrostatic equilibrium.

2.1 Schematic Stellar Evolution

The overall picture is that of a star contracting as it radiates energy outward while its central density and temperature increase. (Thus we sometimes speak of stars as having "negative heat capacity," meaning that they get hotter while losing energy. The explanation of this phenomenon is that in the contraction phase the energy source — i.e., both the radiated and the internal thermal energy — is the gravitational energy released during the contraction.) The process of contraction, radiation, and heating could continue almost endlessly were it not for the nuclear reactions that begin to take place at a certain point along the evolutionary track.

Nuclear reactions set in when the temperature at the stellar centre reaches the threshold for nuclear activity, and the energy liberated in these reactions balances the energy deficit resulting from radiation. The first nuclear reaction is the fusion of hydrogen into helium, whose temperature threshold is about eight million degrees. With the nuclear reactions taking place, a steady-state of equilibrium is formed in which the pressure gradient created by the density and the temperature balances the gravitational force. The energy liberated in the nuclear reactions balances the energy lost by radiation. This equilibrium lasts as long as there is nuclear "fuel" at the centre of the star. The amount of energy liberated in hydrogen fusion is huge: 6×10^{18} erg per each gm of hydrogen converted into helium. The luminosity of a star like the Sun is about 3.9×10^{33} erg per sec. This means that about 6×10^{14} gm of hydrogen, which is about 3×10^{-19} of the solar mass, converts to helium each second. Assuming that the central tenth of the Sun can be burnt to helium, this phase can last for about 3×10^{17} sec, or about 10^{10} yr. This phase is called the *main sequence phase*.

When the nuclear fuel (hydrogen) at the centre of the star is consumed, the nuclear reactions stop and the contraction continues. With contraction, heating persists as well, until the temperature reaches 80 to 90 million degrees, the threshold for helium fusion into carbon and oxygen. A steady-state of equilibrium is again achieved, until the helium at the centre is consumed. The contraction then continues, with increasing temperature, until the star reaches the threshold for the next reaction — the fusion of carbon at a temperature of about 500 million degrees.

We can draw a schematic figure of this evolutionary track. If we choose the central density (ρ_c) to characterize the state of the star and draw its evolution as a function of time, we obtain fig. 2.2.

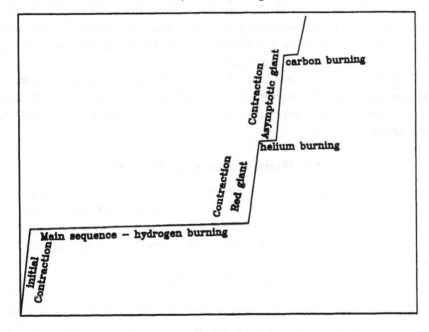

Figure 2.2. Central density in a star vs. time.

In this figure, which is not drawn to scale, the inclined segments represent the contraction phases, when the star radiates, contracts, and heats up. The nearly horizontal segments represent the steady-state phases, when the energy produced by nuclear reactions replenishes energy losses by radiation. We shall later consider in detail the questions of up to what point such an evolution can be followed, and what the parameters are that determine the evolutionary rate. Note that the general trend is of an increase in the central density of the star. There are special cases in which the central density decreases. Usually when the central density increases, the radius of the star also increases owing to a decrease of the density in the outer part of the star.

Let us now turn to the physical processes that determine the structure of stars. For the sake of simplicity, we shall deal with spherically symmetric systems. This means that we shall consider only the variables dependent on the distance from the centre. We will ignore effects which do not possess spherical symmetry, such as those connected with rotation and magnetic fields. In a spherically symmetric system, we deal with concentric mass shells whose depth is dr and surface area is $4\pi r^2$. The volume enclosed in such a shell is $dV = 4\pi r^2 dr$. The structure of the star is determined by the physical laws which govern the processes taking place inside the star and by the initial parameters (the stellar mass and composition) determining the rate of these physical processes.

In dealing with the variables which describe stellar structure, we shall assume that a star has a given mass and initial composition. The influence of the mass and the composition on the stellar physics will be discussed separately.

2.2 Hydrostatic Equilibrium

As mentioned above, there are two main forces acting inside a star: the gravitational force, which acts inward and drives a contraction; and the force created by the pressure gradient, which acts in the direction opposite to gravity. We write Newton's equation of motion for a volume element in the star:

$$\rho\ddot{r} = -\frac{dP}{dr} - \frac{Gm(r)\rho}{r^2} \tag{2.1}$$

where ρ is the density, $\ddot{r}$ is the second derivative of r with respect to time, P is the pressure, r is the radius (the distance from the centre), $m(r)$ is the mass enclosed in a sphere whose radius is r, and G is the Newtonian gravitational constant. The minus sign of the first term on the right-hand side signifies that the force created by the pressure gradient acts in the direction of decreasing pressure. The minus sign of the second term means that gravity acts toward the centre. This equation holds for each volume element, at each point in the star. In a state of equilibrium, the left-hand side of the equation vanishes, and eq. 2.1 takes the form:

$$\frac{dP}{dr} = -\frac{Gm(r)\rho}{r^2}.$$
(2.2)

This is the equation for hydrostatic equilibrium. To get a sense of the magnitude of the time scale governing the process connected with this equation, we might consider what happens when the term $\frac{dP}{dr}$ vanishes in eq. 2.1, and the stellar matter free-falls toward the centre with an acceleration, $-\frac{Gm}{r^2}$. Let us calculate how much time is needed for a mass element at the stellar surface to fall to the centre, according to the kinematic formula $l = \frac{1}{2}\ddot{r}t^2$. We get:

$$t = \sqrt{\frac{2R^3}{GM}}$$
(2.3)

where R is the stellar radius and M is its mass. When we substitute for R and M their values for the Sun, we find $t \simeq 2200$ seconds. This means that the time scale for dynamic processes (in the absence of hydrostatic equilibrium) is of the order of thousands of seconds. This time scale is called the *dynamic time scale*. We have already seen that the other time scales relevant to stellar evolution — i.e., the thermal and the nuclear time scales — are much longer. Hence we assume that a star reaches hydrostatic equilibrium very quickly. Except in extreme cases, such as a sudden collapse or explosion, stars exist for most of their lives in a state of hydrostatic equlibrium. The Newtonian equation of motion thus assumes its form as in eq. 2.2.

Equation 2.2 is a differential equation for the pressure inside a star as a function of the radius. Using an approximate integration of this equation we can get a sense of the conditions in the interior of a star. Assuming that the density is constant, we can substitute $m(r) = \frac{4\pi}{3}r^3\rho$ for the mass, and obtain:

$$dP = -\frac{4\pi}{3}G\rho^2 r dr.$$
(2.4)

We can integrate this equation from the surface to the centre. Recalling that on the surface the pressure practically vanishes and $r = 0$ at the centre, we find for the central pressure, P_c:

$$P_c = \frac{4\pi}{6}G\rho^2 R^2 = \frac{3}{8\pi}\frac{GM^2}{R^4}.$$
(2.5)

To achieve the last expression in eq. 2.5, we have substituted: $\frac{4\pi}{3}\rho R^3 = M$. Substituting the relevant M and R for the Sun, we find $P_c = 1.3 \times 10^{15}$ dyn cm$^{-2} \simeq 10^9$ atm. Substituting this value of P_c in the equation of state for an ideal gas, using for the density the average density of the Sun, we find $T_c \simeq 10^7$ degrees. Although the calculations carried out here are rather crude, we nevertheless obtain an estimate for the range of temperatures relevant to the internal structure of stars. At such a temperature matter exists in a gaseous state and is completely ionized. The matter at the stellar interior is in a state of plasma. The temperature is above the threshold for hydrogen fusion, and nuclear reactions take place at the centre of a star on the order of magnitude of the Sun.

Equation 2.2 includes other variables in addition to the pressure: the mass m and the density ρ. There is another equation which couples the mass and the density. This is the equation of mass continuity:

$$\frac{dm}{dr} = 4\pi\rho r^2. \tag{2.6}$$

By rearranging eq. 2.2, differentiating it again with respect to r, and substituting $\frac{dm}{dr}$ from eq. 2.6, we get:

$$P'' + P'\left(\frac{2}{r} - \frac{\rho'}{\rho}\right) = -4\pi G\rho^2 \tag{2.7}$$

where $P' = \frac{dP}{dr}$. If we substitute the adiabatic equation of state which relates the pressure to the density in eq. 2.7, we can solve it analytically. We shall return to this topic later.

The equation of hydrostatic equilibrium imposes a condition on the pressure gradient. A given pressure, however, can be created by various combinations of densities and temperatures, and the condition on the pressure gradient alone cannot furnish a unique model of a star. For this purpose we need to have a condition on another thermodynamic variable, which together with the condition on the pressure gradient, will yield a unique model of a star. This condition is supplied by the equation describing the heat transport in a star.

2.3 The Equation of Heat Transport

The star loses energy by radiating it from the stellar surface to the surrounding space. The lost energy is replenished by the production of nuclear energy at the centre of the star. Hence the stellar structure should enable energy transport from the centre to the surface.

There are several mechanisms that facilitate energy transport: radiation, convection, conduction, and neutrino radiation. We shall discuss here only radiation and convection, which are the two main mechanisms active in most stars. The other mechanisms will be considered in connection with special conditions for which their contribution is significant. Heat flows down the temperature gradient. The equation for heat transport is thus an equation for the temperature gradient and furnishes the condition on a second thermodynamic variable determining the stellar structure.

Radiation

The rate of energy transport by radiation depends on the temperature gradient and on the transparency (or conversely, on the opacity) of the matter through which the energy flows. The equation for radiative transport may be obtained from the equation for the force exerted on matter by radiation pressure. The radiation pressure, P_{rad}, is given by $P_{rad} = \frac{1}{3}aT^4$, where a is a constant whose value is: $a = 4\sigma/c = 7.56 \times 10^{-15}$ erg cm^{-3} T^{-4}. The force per unit volume exerted by radiation pressure , F_{rad}, is equal to the gradient of the radiation pressure, with a minus sign:

$$F_{rad} = -\frac{dP_{rad}}{dr} = -\frac{4}{3}aT^3\frac{dT}{dr}. \qquad (2.8)$$

On the other hand, the force exerted by the radiation pressure on the matter through which it passes equals the rate at which this matter absorbs the momentum of the radiation. This rate is given by the momentum density of the radiation flux H/c (where H is the energy flux of the radiation and c is the velocity of light) multiplied by a factor which shows what fraction of the radiation's momentum is absorbed in the matter. This factor is given by the product $\kappa\rho$, where κ is the absorption coefficient per unit mass. Such treatment is correct in the approximation used to calculate the radiative flux, called the *diffusion approximation*. In this treatment we consider the radiative flux as a diffusion of photons along the temperature gradient, from hotter to cooler regions. This method is considered a good approximation when the mean free path of the photons (given by $1/\kappa\rho$) is small relative to the temperature scale height, l_T, given by:

$$\frac{1}{l_T} = \frac{1}{T}\frac{dT}{dr}.$$

This condition is fulfilled in most stellar interiors. Using eq. 2.8 we get:

$$-\frac{4}{3}aT^3\frac{dT}{dr} = \kappa\rho\frac{H}{c}. \qquad (2.9)$$

Rearranging eq. 2.9 and substituting $L = 4\pi r^2 H$, where $L(r)$ is the luminosity (the amount of energy radiated from a surface of a sphere of radius r per unit time), we obtain:

$$\frac{dT}{dr} = -\frac{3\kappa\rho L}{4ac4\pi r^2 T^3}. \tag{2.10}$$

This is a differential equation for the temperature. From this equation it is clear that the demand on the temperature gradient increases with the opacity. Solving eq. 2.10 for a given luminosity provides the temperature profile in the star.

We can use eqs. 2.8 and 2.9 in order to calculate, for a given luminosity and opacity, when the force exerted by the radiation pressure will overcome the gravitational attraction. We insert the right-hand side expression of eq. 2.9 into the equation of the hydrostatic equilibrium and substitute $H = L/4\pi r^2$ to find:

$$\frac{dP_{rad}}{dr} = -\frac{\kappa\rho L}{4\pi r^2 c} = -\frac{GM\rho}{r^2} \tag{2.11}$$

where the right-hand side of this equation is adopted from eq. 2.2.

Rearranging this equation we get:

$$L_{Edd} = \frac{4\pi cGM}{\kappa} \tag{2.12}$$

where L_{Edd} is the *Eddington luminosity,*[1] the luminosity whose radiation pressure overcomes the gravitational attraction. Thus the maximal luminosity that can be radiated by a star without tearing it apart has this value. It is interesting to note that since both radiation pressure and gravity depend on the expression ρ/r^2, this expression is factored out. As a result, the Eddington luminosity depends only on the stellar mass, M, and the opacity, κ. Since there is always some gas pressure, the limiting luminosity is actually lower than the calculated L_{Edd}. When we substitute numerical values in the expression for L_{Edd}, we find:

$$\frac{L_{Edd}}{L_\odot} \simeq \frac{1.25 \times 10^4}{\kappa} \frac{M}{M_\odot}. \tag{2.13}$$

Equation 2.10 holds for radiative transport. When convection starts, it is usually more efficient than heat transport by radiation. Let us now consider the conditions under which convection develops in a star.

Convection

Whether convection develops depends on how the density of a volume element changes when it is displaced from its location. Consider a small volume cell. If, as a consequence of this movement, a force develops which returns the cell to its original location, the star will be stable against convection. If, however, a movement creates a force which pushes the cell further in its motion, the star will be unstable in this respect. Convection will develop as a consequence. Let us follow a volume cell in a star which, by random motion, moved upward from its original location (see fig. 2.3).

Because of the spherical symmetry of the configuration, we are interested only in vertical motions. As discussed in Section 2.2, the response to pressure differences is very rapid. The pressure inside the volume cell will immediately become equal to the pressure of the cell's new surroundings. Because heat transfer is a relatively slow process, we assume that the cell did not lose energy to its surroundings during its movement. Hence the changes in its interior are adiabatic, and the density inside the cell changed adiabtically.

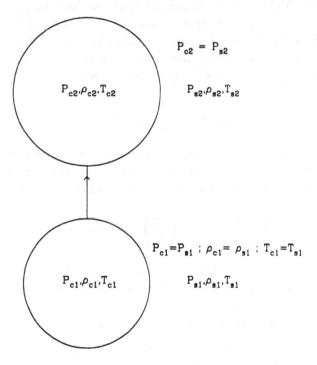

$$P_{c2} = P_{s2}$$

$$P_{c2}, \rho_{c2}, T_{c2} \qquad P_{s2}, \rho_{s2}, T_{s2}$$

$$P_{c1} = P_{s1} \; ; \; \rho_{c1} = \rho_{s1} \; ; \; T_{c1} = T_{s1}$$

$$P_{c1}, \rho_{c1}, T_{c1} \qquad P_{s1}, \rho_{s1}, T_{s1}$$

Figure 2.3. The forces on a convective cell. Subscripts c denote quantities taken in the convective cell, and subscripts s denote quantities taken in the surrounding of the cell. Subscripts 1 denote quantities taken at the initial location of the cell, and subscripts 2 denote quantities taken at the new location, where the cell was moved by a random motion.

How will the ratio between the density inside the cell and the density of its surroundings affect stability?

Consider a volume cell displaced a small distance outward (upward). If the new density inside the cell is higher than the density of the surroundings, then the gravitational force on the cell will be greater than the force on its surroundings. The cell, as a result, will be pulled back to its original location by gravity, and the matter in this environment will be stable against convection. If the new density inside the cell is lower than the density of the surroundings, a buoyant force will act on the cell, pushing it further in the direction of its motion. Here, convection will develop. The density gradient is negative — that is, on moving outward (upward) the density decreases. Thus if the adiabatic change in the density of a cell is greater than the change in the density of the surroundings along its path, the cell will become "lighter" than its surroundings, and will be pushed further along its course. This means that for a star to be stable against convection, the change in its density proceeding outward in a radial direction should be greater than the adiabatic change. The change in the density due to a movement along a distance δr is $\frac{d\rho}{dr}\delta r$, and we get:

$$|\frac{d\rho}{dr}|_{star}\delta r > |\frac{d\rho}{dr}|_{ad}\delta r \implies stable \qquad (2.14)$$

where the subscripts $_{star}$ and $_{ad}$ denote the stellar and the adiabatic quantities respectively. We consider absolute values of the gradients.

Since the density gradient is negative, eq. 2.14 can be written:

$$\left(\frac{d\rho}{dr}\right)_{star}\delta r < \left(\frac{d\rho}{dr}\right)_{ad}\delta r \implies stable. \qquad (2.15)$$

We can convert this condition to an equation for the temperature gradient by using the equation of state for an ideal gas. We must assume that no change in the composition or in the ionization takes place in this zone. From the equation of state for ideal gas, we have: $\frac{dP}{P} = \frac{d\rho}{\rho} + \frac{dT}{T}$. Substituting for $d\rho_{star}$ in eq. 2.15 yields (noting that $\frac{dP}{dr}_{star} = \frac{dP}{dr}_{ad}$):

$$\frac{1}{T}\left(\frac{dT}{dr}\right)_{star}\delta r > \left(\frac{\gamma-1}{\gamma}\right)\frac{1}{P}\left(\frac{dP}{dr}\right)\delta r \implies stable. \qquad (2.16)$$

where γ is the adiabatic exponent (the ratio of the heat capacity at constant pressure to the heat capacity at constant volume).

Now, $\frac{T}{P}\left(\frac{\gamma-1}{\gamma}\right) dP = dT_{ad}$, and eq. 2.16 reads:

$$\left(\frac{dT}{dr}\right)_{star} \delta r > \left(\frac{dT}{dr}\right)_{ad} \delta r \implies \text{stable} \qquad (2.17)$$

and in absolute values we have:

$$\left|\frac{dT}{dr}\right|_{star} < \left|\frac{dT}{dr}\right|_{ad} \implies \text{stable}. \qquad (2.18)$$

What this equation means is that for a star to be stable against convection, its temperature gradient must be more moderate than the adiabatic temperature gradient. If the temperature gradient for radiative transfer demanded by eq. 2.10 is more moderate than the adiabatic temperature gradient, all the energy will be transferred by radiation, and no convection will develop. However, if the temperature gradient demanded by eq. 2.10 is steeper than the adiabatic gradient, then convection will start. Most of the energy will be carried by convection, which is more efficient than radiative transport. The test for convection formulated in eq. 2.18 is a local test, meaning that it should be carried out at each point along the stellar radius. At any point where this test for stability fails, convection starts. This criterion for stability against convection is called the *Schwarzschild criterion* after Karl Schwarzschild who proposed it in 1906. Equation 2.10 shows that for a given luminosity, the demand on the temperature gradient increases with the opacity. Thus when the opacity in the star increases, the temperature gradient becomes ever steeper, though not without limit. When the temperature gradient becomes steeper than the adiabatic gradient, convection begins — a process upon which the opacity has no influence.

Convection transfers heat down the temperature gradient as well. In a convective zone, energy transport by radiation still takes place. The total luminosity transferred equals the sum of the radiative and the convective luminosities:

$$L = L_{rad} + L_{conv} \qquad (2.19)$$

where L, L_{rad}, and L_{conv} are the total luminosity, the radiative luminosity, and the convective luminosity respectively. Equation 2.10 can be rearranged into an equation for the radiative luminosity at a given temperature gradient:

$$L_{rad} = -\frac{4ac4\pi r^2 T^3}{3\kappa\rho}\frac{dT}{dr}. \qquad (2.20)$$

We now have to find an expression for the convective luminosity. Convection is, as yet, not well understood. In dealing with this topic we rely on arguments of a dimensional character. Here, there is at least one free parameter whose value is unknown and it is chosen arbitrarily.

The convective heat transport is calculated by assuming that a matter element (called a convective cell) moves along a certain distance, then dissolves into the surrounding material. This distance is called the *mixing length*. The theory described here, which was developed by E. Bohm-Vittense[2] in 1958, is called the *mixing length theory*. According to eq. 2.18, convection develops when the change in the temperature inside an upward moving convective cell is lower than the change in its surroundings. This means that the temperature of the convective cell when it dissolves is higher than the temperature of its surroundings, and the cell will transfer its excess energy to the environment. There is, as yet, no theory which can determine the value of the mixing length from first principles. This is the free parameter which is to be chosen arbitrarily in the theory, within a range of a few times the pressure scale height as defined below. We shall see later that the value of this parameter has a significant influence on the efficiency of convective transport. Usually this value is assigned the order of magnitude of the pressure scale height: the distance along which the pressure falls by a significant fraction of its original value. We define the pressure scale height, l_P, by:

$$\frac{1}{l_P} = \frac{1}{P}\frac{dP}{dr} = \frac{dlnP}{dr}. \tag{2.21}$$

The convective flux equals the temperature difference between the cell and its environment, times the heat capacity (at constant pressure) of the matter, times the matter density, times the average velocity of the convective cell along the mixing length. The convective luminosity equals the convective flux times $4\pi r^2$. Thus,

$$L_{conv} = 4\pi r^2 \bar{v}\rho c_p \left(|\frac{dT}{dr}|_{star} - |\frac{dT}{dr}|_{ad} \right) l \tag{2.22}$$

where $\bar{v}$ is the average velocity, c_p is the heat capacity at constant pressure, and l is the mixing length. We have to calculate the average velocity and determine l, the mixing length.

We calculate $\bar{v}$ by using the Newtonian equation $\bar{v} = \sqrt{2al}$, where a is the acceleration along the path. This acceleration is found by calculating the buoyant force per unit volume acting on the convective cell. This variable equals the average difference between the cell density and its environment density, times the gravitational acceleration. We write:

$$F = \frac{1}{2}g \left(\left|\frac{d\rho}{dr}\right|_{ad} - \left|\frac{d\rho}{dr}\right|_{star} \right) l \qquad (2.23)$$

where g is the gravitational acceleration. The factor $\frac{1}{2}$ is introduced in order to obtain the average between the maximum acceleration and the minimal acceleration (which is zero). The acceleration is given by:

$$a = \frac{F}{\rho} = \frac{1}{2}\frac{g}{\rho} \left(\left|\frac{d\rho}{dr}\right|_{ad} - \left|\frac{d\rho}{dr}\right|_{star} \right) l. \qquad (2.24)$$

Substituting in eq. 2.22: $\bar{v} = \sqrt{2al}$ and

$$\left(\left|\frac{d\rho}{dr}\right|_{ad} - \left|\frac{d\rho}{dr}\right|_{star} \right) = \frac{\rho}{T} \left(\left|\frac{dT}{dr}\right|_{star} - \left|\frac{dT}{dr}\right|_{ad} \right)$$

we find:

$$L_{conv} = 4\pi r^2 \rho c_p \sqrt{\frac{g}{T}} \left(\left|\frac{dT}{dr}\right|_{star} - \left|\frac{dT}{dr}\right|_{ad} \right)^{3/2} l^2. \qquad (2.25)$$

We can still use the equation of hydrostatic equilibrium, where we have $g = \frac{1}{\rho}\frac{dP}{dr}$, and $l = \alpha l_P$ (where l_P is the pressure scale height and α is a dimensionless constant of the order of unity), to get:

$$L_{conv} = 4\pi r^2 \rho c_p \sqrt{\frac{1}{\rho T}\frac{dP}{dr}} \left(\left|\frac{dT}{dr}\right|_{star} - \left|\frac{dT}{dr}\right|_{ad} \right)^{3/2} \alpha^2 l_P^2. \qquad (2.26)$$

It is evident from this equation that the value of α has a significant influence on the value of L_{conv}. In the literature we can find values for α from 0.5 to 2.5.

We have been dealing here with random motions of the convective cell in the outward direction. This was done only for the sake of simplicity. All the considerations remain the same for inward random motions, with the opposite signs appearing wherever we deal with direction-dependent variables. The inward moving cells are cooler than their environment when they dissolve, and they cool the inner parts of the star. The net result is the same: convection, when it exists, transfers heat outward.

Our treatment of convection in the preceding section was somewhat simplified. We assumed linear motion of the mass elements without friction. Obviously the situation in nature is more complicated, when friction (or viscosity) and mutual interaction between opposite directed streams cause a loss of energy and the formation of turbulent motions. This subject is currently under intensive study. In the coming years more sophisticated methods will likely be used to treat convection in stars.

Equation 2.26 can also be considered as an equation for the temperature gradient in convective zones. In most of the stellar interior, while convection is operating and the density is sufficiently high, the difference between the stellar temperature gradient and the adiabatic temperature gradient (needed to transfer the energy by convection) is very small — less than one-thousandth of the value of the temperature gradient. Hence the temperature gradient in the star in such regions can be regarded as the adiabatic temperature gradient. In the outermost regions of stars, where the density drops to very low values, the convective flux is less efficient. The stellar temperature gradient demanded by eq. 2.26 is much steeper than the adiabatic one. We call such regions *super-adiabatic zones*.

2.4 The Heat Equation

Conservation of energy must be maintained throughout all the processes which take place in stars. We write this in a form of an equation for each volume element:

$$\frac{d\varepsilon}{dt} = \rho q - \frac{1}{4\pi r^2}\frac{dL}{dr} - \frac{P}{V}\frac{dV}{dt} \tag{2.27}$$

where ε is the internal energy per unit volume, q is the rate of nuclear energy generation per unit mass per second, V is the volume, and L is the luminosity. The left-hand side of this equation shows the energy gain of a unit volume per unit time. The first term on the right-hand side of the equation is the amount of nuclear energy produced per unit volume per unit time. The term $\frac{dL}{dr}$ shows the difference between the luminosity which entered the unit volume from its inner neighbour and the luminosity which left this unit volume to its outer neighbour. The last term on the right-hand side shows the amount of energy lost by work in expanding. This equation should be satisfied at each point in the star. In static conditions we expect the time derivatives to vanish and are left with:

$$\frac{dL}{dr} = 4\pi r^2 \rho q. \tag{2.28}$$

In those regions in the star where the temperature is below the threshold for nuclear activity, the right-hand side of eq. 2.28 vanishes and we have: $\frac{dL}{dr} = 0$.

2.5 The Full Set of Equations

Together we have four differential equations to determine the structure of a star:

$$\frac{dP}{dr} = -\frac{Gm(r)\rho}{r^2} \qquad (2.29.1)$$

$$\frac{dm}{dr} = 4\pi r^2 \rho \qquad (2.29.2)$$

$$\frac{dL}{dr} = 4\pi r^2 \rho q \qquad (2.29.3)$$

and

$$\frac{dT}{dr} = \frac{-3\kappa\rho L}{4ac4\pi r^2 T^3} \qquad \text{for a radiative zone,} \qquad (2.29.4a)$$

or

$$\frac{dT}{dr} = \left(\frac{\gamma - 1}{\gamma}\right)\frac{T}{P}\frac{dP}{dr} \qquad \text{for a convective zone.} \qquad (2.29.4b)$$

In eq. 2.29.4b we used the adiabatic temperature gradient for convective zones, assuming that in most situations this equation describes to a very good approximation the conditions in convective zones.

We have seven dependent variables in these equations: mass, density, temperature, pressure, opacity, luminosity, and rate of the nuclear reactions. There are also the three equations of the gas characteristics: the equation of state which relates pressure to temperature and density; the equation which gives opacity as a function of temperature and density; and the equation which gives the rate of the nuclear reactions as a function of temperature and density. We thus have seven equations for the seven variables. Four of these are differential equations, and the remaining three are algebraic relations.

We have, as well, four boundary conditions for the differential equations. At the centre of the star, mass and luminosity vanish; and at the stellar surface pressure and temperature are practically zero. The equations plus the four boundary conditions determine a unique solution for the structure of a star with a given mass and composition.

The problem is that these equations cannot be solved analytically. Moreover, the dependence of the gas characteristics on the thermodynamic variables vary over wide ranges in different regions of the star.

Until the 1960s, the common way of solving the equations for the stellar structure was to make certain assumptions about the dependence of the gas characteristics on the thermodynamic variables in different regions of the star (such as the central region, the envelope, etc.). The differential equations were solved analytically in these regions. Demanding continuity in the functions and their derivatives at the boundaries between these regions allows the attainment of more realistic solutions. More details about this method are offered in Schwarzschild's book.[3]

With the appearance in the late 1960s and 1970s of big and fast computers having very large computational power, numerical methods have been developed in which the values of the gas characteristics are introduced into the computation scheme at each point in the star. The equations are then solved numerically, yielding the detailed structure of the star. We shall later deal in greater detail with these methods.

References

1. Eddington A.S., 1930, in *The Internal Constitution of the Stars*, Cambridge University Press, Cambridge.
2. Bohm-Vittense E., 1958, *Zs. f. Astrophys.*, **46**, 108.
3. Schwarzschild M., 1958, in *Structure and Evolution of Stars*, Princeton University Press, Princeton.

Chapter 3

The Gas Characteristics

In the preceding chapter we discussed the differential equations determining the structure of a star. To complete the set of equations needed to solve for the seven variables appearing in these equations, we have to add the three equations for the gas characteristics: the equation of state, the equation for opacity, and the equation for the rate of nuclear reactions.

3.1 The Equation of State

The thermodynamic state of a system with a given chemical composition can be fully determined when supplied with two thermodynamic variables. These variables can be any two of a long list of thermodynamic variables: pressure, temperature, density, energy, entropy, and so on.

The equation of state is an equation relating three of the variables to one another so that any one of them can be determined by the other two. We shall now treat the most common form, which gives pressure as a function of density and temperature. Many other combinations of the equation of state can be used. We shall mention some of them when we deal with the solutions for stellar structure.

To calculate the pressure in gaseous matter, let us consider a volume of gas surrounded by solid walls. The pressure is the force per unit area which the gas exerts on the surrounding walls, found by summing up the contributions of all the particles to this force. We find the contribution of each particle by calculating the amount of momentum exchanged by the particle in an elastic collision with a solid object. The sum of all these momenta exchanges per unit time per unit area of the solid object is the pressure exerted on this object. Obviously, because of the huge number of particles and their statistical behaviour, the summation is performed by integration. Without going into details, we present here the expression for the pressure in the form of an integral:

$$P = \frac{1}{3} \int_0^\infty \vec{p} \cdot \vec{v} \, n(\vec{p}) d\vec{p} \qquad (3.1)$$

where $n(\vec{p})$ is the distribution function which represents the distribution of the particles between the different momentum states and $d\vec{p}$ is the three-dimensional differential of the momentum. The terms for pressure in the integrand are the product of a particle's momentum $\vec{p}$ times its velocity $\vec{v}$ (whose integration yields the amount of momentum per unit time reaching the solid walls) and times the distribution function. The integration is carried out over the whole range of the momentum space, from zero to infinity.

3.2 Ideal Gas

The simplest form of the equation of state is that for an ideal gas. We define an ideal gas as an assembly of particles subject to random motions, where elastic collisions are the only interactions between the particles. The particles can exchange momentum and kinetic energy through the collisions, but these collisions are restricted to the conservation laws of kinetic energy and momentum. The kinetic energy of a particle, within the classical limit, ϵ_k, is given by the Newtonian expression $\epsilon_k = \frac{1}{2}mv^2$, where m is the particle mass and v its velocity. The momentum $\vec{p}$ is given by $\vec{p} = m\vec{v}$. We find that the product of the momentum times the velocity, which appears in the integrand of the pressure, equals:

$$\vec{p} \cdot \vec{v} = mv^2 = 2\epsilon_k. \qquad (3.2)$$

The distribution of the particles represented by $n(\vec{p})$ is the Boltzmann distribution. In this simple case we find a relation between the pressure and the kinetic energy of the gas:

$$P = \frac{2}{3}\epsilon \qquad (3.3)$$

where ϵ is the kinetic energy of the gas per unit volume. The integration for the pressure yields:

$$P = nkT \qquad (3.4)$$

where n is the number density of the particles (number of particles per unit volume), and k is the Boltzmann constant. At the temperatures existing in stars, the gas is monoatomic and practically fully ionized. The number of particles in a sample of gas equals the number of ionized atoms plus the number of free electrons. The number density of particles is related to the mass density ρ by:

$$\rho = n\bar{m} = n\mu m_p \tag{3.5}$$

where $\bar{m}$ is the average mass of the particles, m_p is the proton mass, and μ is a coefficient which shows the ratio of the average mass of the particles to the proton mass. For example, in a monoatomic neutral hydrogen gas, $\mu = 1$ because the mass of each particle equals (approximately) the proton mass. If this gas is fully ionized, for each atom there are two particles: the hydrogen nucleus and the electron. For this gas $\mu = \frac{1}{2}$. In a fully ionized gas which contains various elements we can write:

$$\frac{1}{\mu} = 2X + \frac{3}{4}Y + \frac{1}{2}Z \tag{3.6}$$

where X and Y are the fractions by mass of hydrogen and helium respectively, and Z is the fraction of the heavier elements. These elements are usually called metals. $\frac{1}{\mu}$ is the number of particles per proton mass.

We derive the coefficient $\frac{3}{4}$ of Y in eq. 3.6 from the fact that ionized helium accounts for three particles (a nucleus and two electrons), and its mass is four times the proton mass. We obtain the coefficient $\frac{1}{2}$ of Z from the fact that the number of electrons in a heavy element is about one-half of its atomic mass.

Substituting for n in eq. 3.4, we find for ideal gas in the classical limit:

$$P = \frac{k}{\mu m_p}\rho T. \tag{3.7}$$

This form can be used in the equations which determine the stellar structure, after substituting the appropriate value of μ for the actual conditions in the star. (It is worth noting here that the value of μ needed for calculating the thermodynamic conditions in the star is determined by the same thermodynamic conditions. We shall tackle this problem again and again. The actual solution for the structure of a star is obtained by a long process of iteration.)

The state of an ideal gas in the classical limit exists for sufficiently low densities and for medium to high temperatures. When we come to deal with a range of very high temperatures, when the thermal velocities of the particles exceed a few percent of the speed of light, we have to consider the relativistic equation of state. On the other hand, when we come to a range of very high densities, we have to use the equation of state for degenerate matter.

3.3 Relativistic Regime for Ideal Gas

When the thermodynamic velocities become relativistic, the product $\vec{p} \cdot \vec{v}$ is not related linearly to the kinetic energy of the particles. We have:

$$\vec{p} \cdot \vec{v} = mv^2 = \frac{m_0 v^2}{\sqrt{1 - \frac{v^2}{c^2}}} \tag{3.8}$$

where m_0 is the rest mass of the particle. The expression for the kinetic energy is:

$$(m - m_0)c^2 = m_0 \left(\frac{1}{\sqrt{1 - \frac{v^2}{c^2}}} - 1 \right) c^2. \tag{3.9}$$

To simplify matters, let us consider a photon gas. For photons the rest mass is zero, the velocity is c, and their energy is kinetic energy, ϵ. The product $\vec{p} \cdot \vec{v} = \vec{p} \cdot \vec{c} = \epsilon$, and from the integration of eq. 3.1, we find for the relation between the pressure and the (kinetic) energy:

$$P = \frac{1}{3}\epsilon = \frac{1}{3}aT^4 \tag{3.10}$$

where $a = 4\sigma/c$ is a constant. We have already used this expression in Section 2.3 of Chapter 2. In an actual situation the state is not fully but only partially relativistic. We use an equation which approximates the pressure in a semirelativistic state.

The particles in the gas are free atomic nuclei and free electrons. The average energy of the nuclei is the same as the average energy of the electrons. Because of the high ratio of their masses, the average velocities of the nuclei are much lower than the average velocities of the electrons. Thus, when the electron gas is already relativistic, the nuclei gas may still behave like an ideal gas in the classical limit, and we have to calculate separately the contributions to the pressure of the two kinds of particles.

3.4 Degenerate Gas

At very high densities, Pauli's exclusion principle begins to play an important role. According to this principle, which is valid for particles that possess a half-integer spin, no two particles can occupy the same quantum state. Particles with a half-integer spin are called *fermions*. A quantum state for a free fermion is defined as a cell of volume h^3 (where h is the Planck constant) in the six-dimensional phase space defined by the product of the regular spatial space (x, y, z) times the momentum space (p_x, p_y, p_z).

Actually, each cell of this type can accommodate two particles with opposite spins because a different spin state means a different quantum state. The number density of particles is the reciprocal of the volume of the spatial cell occupied by each particle. An increase in the number density is a decrease of the spatial component of the particle cell in phase space. This decrease demands an increase in the momentum component of the cell. Thus in high density matter the momentum cells are larger, the number of states available at each momentum level becomes smaller, and the particles are forced to occupy higher momentum levels. A degenerate state is one in which the particles occupy momentum levels that are much higher than expected from the temperature of the gas.

We calculate the pressure of the gas by using the momentum of the particles. In a degenerate gas, the lower momentum levels are full, and the particles which determine the gas properties are those which occupy the uppermost levels. Hence the pressure of a degenerate gas is determined by the momentum of the particles with the highest momentum, irrespective of the temperature of the gas. The term which changes significantly in the integrand of the pressure is $n(\vec{p})$, the distribution function. For low momentum levels its value is unity, which means that the lower levels are fully occupied up to a certain level called the *Fermi level*. Above this level $n(\vec{p}) = 0$, which means that these levels are unoccupied.

We should distinguish again between the nuclei gas and the electron gas. At the same average energy, a nucleus has a higher momentum than an electron by a factor equaling the square root of their mass ratio. The lowest momentum levels of the electrons are therefore lower than those of the nuclei, and at a given density the electrons reach the limit of degeneracy earlier. It might happen that while the electron gas is already degenerate, the nuclei gas still behaves like an ideal gas in the classical limit. We calculate the contribution of each component of the gas separately, and add the partial contributions of the components to yield the total gas pressure.

Substituting the values of $n(\vec{p})$ in the pressure integral, we find that in a fully degenerate (nonrelativistic) gas the pressure is proportional to $\rho^{5/3}$. For a fully degenerate, extremely relativistic gas, the pressure is proportional to $\rho^{4/3}$. The dependence on temperature is negligible in both cases. The numerical values, in c.g.s. units, of the expressions for the pressure of degenerate electron gas are:

$$P_e = 0.991 \times 10^{13} \left(\frac{\rho}{\mu_e}\right)^{5/3} \qquad \text{for a nonrelativistic gas} \qquad (3.11)$$

and

$$P_e = 1.231 \times 10^{15} \left(\frac{\rho}{\mu_e}\right)^{4/3} \qquad \text{for a relativistic gas} \qquad (3.12)$$

where μ_e is the mean molecular weight per electron in the gas. This form of the equation of state is the same as that of the adiabatic equation. The calculations can therefore be carried out on the same basis.

A comparison of the expressions for the pressure in the classical limit of an ideal gas and the degenerate state shows that the contributions of the two states are approximately the same when $T^{3/2} = 4 \times 10^7 \, \rho/\mu_e$, where T and ρ are given in their c.g.s. units. Temperatures higher than this value mostly yield the classical limit of ideal gas states. Lower temperatures yield, for the most part, the degenerate state. Making use of this expression can give an estimate as to the degree to which a system is degenerate.

3.5 The Virial Theorem

A productive statement in many-body physics is the *virial theorem*. This is a statistical statement about systems containing mutually interacting particles.

We consider a system of mass point particles, m_i, located at the coordinates $\vec{r}_i$, and possessing momentum $\vec{p}_i$, and we calculate the time derivative of the sum of the multiplications $\vec{p}_i \cdot \vec{r}_i$:

$$\frac{d}{dt} \sum_i \vec{p}_i \cdot \vec{r}_i = \sum_i \frac{d\vec{p}_i}{dt} \cdot \vec{r}_i + \sum_i \vec{p}_i \cdot \frac{d\vec{r}_i}{dt}. \qquad (3.13)$$

The time derivative of a particle's momentum equals the force acting on that particle: $\frac{d\vec{p}_i}{dt} = \vec{f}_i$, and eq. 3.13 yields:

$$\frac{d}{dt} \sum_i \vec{p}_i \cdot \vec{r}_i = \sum_i \vec{f}_i \cdot \vec{r}_i + \sum_i m_i v_i^2. \qquad (3.14)$$

In a nonrelativistic case, the kinetic energy of a particle equals $\frac{1}{2}mv^2$, and the second term on the right-hand side of eq. 3.14 equals twice the total kinetic energy of the system, T. In an equlibrium state, the time derivative given on the left-hand side of eq. 3.14 vanishes, and we have $2T = -\sum_i \vec{f}_i \cdot \vec{r}_i$. The implications of the virial theorem are used widely in thermodynamics. We shall consider here only its use with regard to stellar structure and evolution.

For mutually interacting particles, each term in the sum $\sum_i \vec{f}_i \cdot \vec{r}_i$ includes the force acting on the ith particle by all the other particles m_j, multiplied by the particle coordinate $\vec{r}_i$: $\vec{f}_i \cdot \vec{r}_i = \sum_j \vec{f}_{ij} \cdot \vec{r}_i$. This expression should be summed for all the particles m_i. Thus:

$$\sum_i \vec{f}_i \cdot \vec{r}_i = \sum_i \sum_j \vec{f}_{ij} \cdot \vec{r}_i.$$

For each $\vec{f}_{ij}$ acting on the ith particle, there is a force $\vec{f}_{ji}$, of which m_j is acted upon by m_i, and $\vec{f}_{ji}$ is equal to $\vec{f}_{ij}$ and directed in the opposite direction. By Newton's third law, $\vec{f}_{ij} = -\vec{f}_{ji}$, and $\vec{f}_{ij} \cdot \vec{r}_i + \vec{f}_{ji} \cdot \vec{r}_j = \vec{f}_{ij} \cdot (\vec{r}_i - \vec{r}_j)$. If the system contains an ideal gas, the only mutual interactions (apart from elastic collisions) are gravitational attraction between any pair of particles, and possibly magnetic and electric forces which we ignore here. Thus we have:

$$\sum_i \vec{f}_i \cdot \vec{r}_i = \sum_{pairs} \vec{f}_{ij} \cdot (\vec{r}_i - \vec{r}_j) = - \sum_{pairs} \frac{Gm_i m_j}{|\vec{r}_i - \vec{r}_j|^2} \cdot (\vec{r}_i - \vec{r}_j)$$

$$= - \sum_{pairs} \frac{Gm_i m_j}{|\vec{r}_i - \vec{r}_j|} = \Omega \qquad (3.15)$$

where $\sum_{pairs}$ means that we sum over all pairs of particles. We designate by Ω the gravitational potential energy of the system. Thus we obtain:

$$2T = -\sum_i \vec{f}_i \cdot \vec{r}_i = -\Omega. \qquad (3.16)$$

If there is no bulk motion, and v_i is the thermal velocity of the ith particle, the total kinetic energy of the gas particles is the thermal energy of the system. If the system is a gaseous sphere existing in an equilibrium state and bound by gravity, we find that the thermal energy of the system equals negative one-half of the gravitational potential energy. This result agrees with the schematic evolutionary track described in Chapter 2; which is to say that while a star contracts, half the potential gravitational energy released in the contraction converts to internal thermal energy. The other half radiates away. This circumstance causes the irreversible nature of the evolutionary track. To reverse the evolutionary direction, the star would have to "collect" energy from the interstellar medium. The thermodynamic rules dictate a time arrow in the events along stellar evolution, and this track is unidirectional.

We should keep in mind that the relation given in eq. 3.16 is correct for a classical ideal gas, where the kinetic energy of a particle equals $\frac{1}{2}mv^2$. In different states, such as relativistic or degenerate states, this relation is different. Thus in a fully relativistic gas, like a photon gas, the momentum $p = \frac{h\nu}{c}$, $v = c$, and the last sum on the right-hand side of eq. 3.13 equals $\sum_i \vec{p}_i \cdot \vec{r}_i = \sum_i T_i = T$ (instead of $2T$ in the nonrelativistic case). In a semirelativistic case, the coefficient of the kinetic energy is between one and two.

For the mutual force between the particles, we considered gravity only. Actually there are other forces, such as the Coulomb force, but their influence is usually of a lower order of magnitude than gravity.

The virial theorem is discussed here with regard to a discrete n-body system. Extending the validity of this statement to systems with continuous variables is justified with appropriate boundary conditions, as in the vanishing of pressure on the surface boundary of a star.

3.6 Opacity

We have seen in Section 2.2 that opacity has a significant role in determining the temperature gradient required to transfer a given luminosity by radiation. When radiation passes through matter, the matter absorbs a certain fraction of the radiation flux. The absorption causes the absorbing particle to ascend to a higher energy level. Usually it re-emits this energy when it descends to its ground level. The magnitude of the fraction absorbed from the radiation flux depends on the atomic processes which take place in the matter's particles. When the matter's particles are un-ionized atoms, the absorption of a photon is followed by the atom's ascension to a higher energy level, or by its ionization. The first process is called a bound-bound (b-b) process; the second is called a bound-free (b-f) process, since the electron moves by this process from a bound state to a free state. When the matter is fully ionized, it includes free ions and free electrons. In this case the absorption of a photon by a free electron in the presence of a heavy ion is called a free-free (f-f) process, for the reason that the free electron moves from one free state to another. A fourth process is that of Thomson scattering, which is actually an elastic collision of a photon with an electron. In this collision no energy is transferred, but the photon is deflected from the directed radiation flux, so that this process also contributes to the decrease of the radiation flux.

Let us now briefly describe these processes. We have already stated that at the temperatures existing in most of the stellar interior, the light elements of hydrogen and helium are fully ionized. They cannot therefore participate in the b-b and b-f processes. The occurrence of these processes depends on the presence of metals, elements which are heavier than helium, in the star. The ionization energies of inner shell electrons of metals are on the order of thousands eV, which is equivalent to temperatures up to few million degrees.

The b-b Process

For a b-b process to occur, the energy of the absorbed photon must be equal to the energy difference between the two energy levels:

$$\epsilon_{ph} = \Delta\epsilon = \epsilon_2 - \epsilon_1 \tag{3.17}$$

where ϵ_{ph} is the photon energy, and ϵ_1 and ϵ_2 are the energies of initial and final states respectively. The probabilities that such transitions will occur are calculated according to the rules of atomic physics. These rules put statistical limitations on the probabilities. These probabilities are larger only at a certain range of temperatures and densities. The calculation of opacity by this procedure is actually a calculation of the opacities in series of absorption lines. No general formula for average opacity can be found by this process. Researchers perform instead detailed calculations of line opacities in certain compositions of matter. The results, in the form of appropriate tables of opacity values for given composition at given temperatures and densities, are fed into the programs which calculate stellar structure.

The b-f Process

This is a process of ionization. The limit on the photon energy in this process is that it should be above the energy needed for ionization. The excess energy over this limit is imparted to the freed electron as kinetic energy:

$$\epsilon_{ph} = \chi + \epsilon_k \tag{3.18}$$

where χ is the ionization energy, and ϵ_k is the kinetic energy of the released electron. This interaction is limited only to photons whose energy is above the ionization energy. The average photon energies are proportional to the local temperature. The maximum probability of this process taking place is at energies which are very close to the above limit. The probability decreases with the increasing photon energy. Thus when calculating a function for the dependence of the absorption coefficient as a function of density and temperature, a special factor is added which reflects the dependence on the specific ionization energy characteristic for each atomic species. This factor is called the *Gaunt factor*. Its form is such that it vanishes up to a certain energy, then jumps to a maximum, from which it decreases gradually with increasing temperature. The order of magnitude of the maximal value of this factor is around unity.

The formula for the the average b-f absorption coefficient, called *Kramers' law of opacity,* is given by:

$$\kappa_{bf} = 4.34 \times 10^{25} \bar{g}_{bf} Z (1 + X) \frac{\rho}{T^{7/2}} \quad \text{cm}^2 \; \text{gm}^{-1} \qquad (3.19)$$

where X and Z are the fraction of the mass of hydrogen and the metals respectively, and $\bar{g}_{bf}$ is the Gaunt factor.

The f-f Process

In an f-f process, a free electron absorbs a photon and jumps from one free state to another. Writing the equations of the conservation of energy and momentum for such a process shows that the two equations cannot be satisfied together for a single electron. Thus for this process to occur, another particle, such as a heavy ion, must participate to absorb the excess momentum in the process. Because of the high ratio between the masses of the ion and the electron, this momentum exchange does not affect the ion significantly. We can still deal with the electron alone, keeping in mind that the conservation rules are satisfied by the participation of a heavy ion.

Using the same approximation methods as for the calculation of κ_{bf}, we find Kramers' law for κ_{ff}:

$$\kappa_{ff} = 3.68 \times 10^{22} \bar{g}_{ff} (X + Y)(1 + X) \frac{\rho}{T^{7/2}} \quad \text{cm}^2 \; \text{gm}^{-1}. \qquad (3.20)$$

Comparing eqs. 3.19 and 3.20, we find that the dependence on the density and temperature is the same in the two equations. The difference between them is in the magnitude of the numerical coefficient, and in the presence of the term $(X + Y)$ in eq. 3.20, whereas eq. 3.19 has the term Z instead. From these differences, and some others not included in our discussion, it is inferred that when Z is about 0.01 the two expressions yield the same value for the opacity coefficients. This means that in the approximations we have used, when Z is greater than 0.01 , the b-f opacity is the dominant of the two. This is the case for stars of the second and third generations. When Z is lower than one percent, as it is for first generation stars, the dominant opacity process is the f-f process.

Electron Scattering

Electron scattering is a process in which no energy is exchanged between the photon and the electron. The photon is "scattered" by the electron; in other words, it changes direction from the directed flux of radiation and is "lost" from the radiation flux. The scattering coefficient does not depend on the temperature nor on the density. It contains the product of a constant factor times the number of electrons per unit mass. In a state of complete ionization we have:

$$\kappa_{es} = \frac{8\pi}{3} \left(\frac{e^2}{m_e c^2} \right)^2 \frac{1}{\mu_e m_p} \quad \text{cm}^2 \text{ gm}^{-1}. \tag{3.21a}$$

Substituting the numerical values, and noting that in an ionized gas $1/\mu_e = (1 + X)/2$, we have:

$$\kappa_{es} = 0.2(1 + X) \quad \text{cm}^2 \text{ gm}^{-1}. \tag{3.21b}$$

This process always takes place, and its contribution should be added to the other processes. At high temperatures and relatively low densities, the contributions of the other processes become very small, and electron scattering becomes the significant process. Comparing eqs. 3.20 and 3.21 we find that the contributions of the f-f process and electron scattering are equal when $\frac{\rho}{T^{7/2}} \sim 5 \times 10^{-24}$. For characteristic densities of 50 to 100 gm cm^{-3}, this equality yields $T \sim 10^7$ degrees. In fig. 3.1 the dominant regions of each kind of opacity are shown in the temperature-density plane.

We find that at very high temperatures ($T > 10^7$ degrees) the dominant process is electron scattering. At the stellar centre the opacity is low (about 0.4 cm^2 gm^{-1}). Moving outward in the star, the temperature and the density decrease; but since the dependence on the temperature is crucial, the opacity increases in this situation reaching values of thousands cm^2 gm^{-1} at temperatures of about 20,000 degrees. At lower temperatures, Kramers' law of opacity no longer holds. When the temperature decreases to below the ionization temperatures of hydrogen and helium, most of the material becomes neutral. The photons' energies are below the ionization or excitation energies of the atoms, free electrons for electron scattering are virtually nonexistent, the interaction of photons with matter becomes very weak, and the matter becomes almost transparent. We shall consider these topics when dealing with the structure of stellar envelopes.

Figure 3.2 shows a graph for the characteristic opacity of a model of a red giant star having one solar mass. Note the steep decline of the opacity

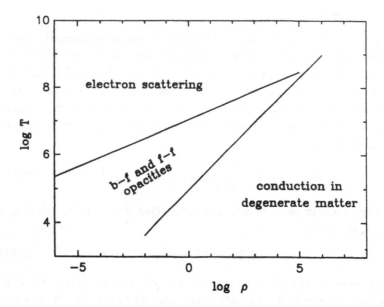

Figure 3.1. Opacities in the density-temperature plane. The boundary between b-f and electron scattering regimes is determined by calculating the density and temperature at which the opacities calculated by eqs. 3.19 and 3.21b are equal. (These values were calculated with $Z = 0.02$, $X = 0.7$.) The boundary between ordinary and degenerate matter is determined by calculating the density and temperature at which the pressure of degenerate electrons equals that of an ideal gas in the classical limit.

curve caused by recombination of hydrogen at the low temperatures at the outer part of the star.

3.7 Rate of Nuclear Reactions

The nuclear reactions which take place in the interior of stars consist of a fusion of light elements to heavier ones, where the excess mass between the initial and the final elements is liberated as thermal energy. This energy replenishes the energy loss from the star by radiation.

The Coulomb barrier, formed by the positive electric charge of the nuclei, prevents the mutual penetration of atomic nuclei which leads to the nuclear transmutations. The particles overcome this barrier by their kinetic energy, whose average value is characterized by the temperature. Since hydrogen atoms have the lowest electric charge, their Coulomb barrier is the lowest.

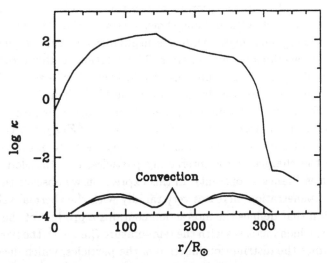

Figure 3.2. Opacity profile of a red giant star of one solar mass. The moderate decrease of the opacity which begins at $R \simeq 140 R_\odot$ is caused by the recombination of helium, while the steep decrease which begins at $R \simeq 280 R_\odot$ is caused by the recombination of hydrogen.

Hydrogen fusion is the first reaction to start when the star heats up at a threshold of about eight million degrees. Heavier elements, which have higher Coulomb barriers, start their fusion at higher temperatures.

The simplest reaction is the fusion of two protons to form deuterium (^{2}D). The nuclear forces are very short-range ones. They become significant only when the interacting particles "penetrate" one another, which means that the distance between their centres is shorter than twice their radii. Within this distance, these forces are larger by orders of magnitude than the electrostatic force. Outside of this radius, however, they are practically zero. We calculate the Coulomb barrier for this process to find that at a distance of 10^{-13} cm between the protons, which is the characteristic "radius" of a proton, the energy barrier, ϵ_{barr}, amounts to:

$$\epsilon_{barr} = \frac{q^2}{r} = 2.3 \times 10^{-6} \text{ erg} \simeq 1.4 \times 10^6 \text{ eV}. \qquad (3.22)$$

This energy is equivalent to a thermal energy of about 10^{10} degrees. The average energy of the protons in the solar centre, for example, is equivalent to 1.5×10^7 degrees, which is far below the Coulomb barrier. We may assume that the energy distribution of the protons is Maxwellian. We can expect that the higher energy protons at the tail end of the distribution are those that interact. Calculations show, however, that the number of particles possessing such high energies is negligible and cannot yield the rate of nuclear reactions we find in stars. The solution to this problem is that the mutual penetration of protons at relatively "low" energies is due to the *tunneling effect* proposed by Gamow in the 1940s. This is a quantum mechanical effect. The probability of penetration occurring is proportional to an expression in the form: $exp[-b/v]$, where $b = 4\pi^2 Z_1 Z_2 e^2/h$ contains the parameters of the barrier, and v is the thermal velocity of the particle (Z_1, Z_2 are the charges of the interacting particles, e is the elementary charge, and h is Planck's constant). In this expression we observe that the probability of penetration increases very rapidly with the thermal velocity.

The function that defines the interaction contains a product of the penetration factor, which increases with the temperature (i.e., with the particles' velocities), times the distribution function of the particles, which decreases with increasing temperature. Calculating this product for a proton-proton reaction at the density of the solar centre, we find that the maximum of this function occurs in the range of 10 to 30 keV, which is equivalent to a temperatures of 1 to 3 $\times 10^8$ degrees.

As we learned in the preceding chapter, the nuclear energy should replenish the energy lost from the star by radiation so as to maintain the temperature needed to create the pressure gradient demanded by the hydrostatic equation. If nuclear energy production exceeds energy losses, the star heats up, expands, and cools down until the energy production is reduced to the appropriate value. If the energy production is below the rate needed to replenish the energy losses, the star cools down, contracts, and heats up until energy production levels with the rate of energy losses. There is a negative feedback mechanism between the rate of nuclear reactions and the thermodynamic variables which maintains the energy balance in the star.

The fusion of hydrogen proceeds in two main channels: the proton-proton (p-p) reaction, and the carbon cycle.

3.8 The p-p Reaction

The first step in the p-p reaction channel is the interaction of two protons to form a deuterium. Upon interacting with another proton, the deuterium forms a ^{3}He isotope. Two ^{3}He nuclei then interact to form a ^{4}He nucleus and two free protons. We sum these steps in tab. 3.1:

Table 3.1.

			energy(MeV)	τ(years)
^{1}H + ^{1}H	$\rightarrow$	^{2}D + e^+ + ν	1.442	8×10^9
^{2}D + ^{1}H	$\rightarrow$	^{3}He + γ	5.493	4.4×10^{-8}
^{3}He + ^{3}He	$\rightarrow$	^{4}He + ^{1}H + ^{1}H	12.859	2.4×10^5

where τ is the characteristic reaction time calculated for the thermodynamic state and the composition at the centre of the Sun, γ, is a photon released in the interaction, ν is a neutrino, and e^+ is a positron. The energy released in each step is displayed in the third column. The positron is annihilated immediately on interacting with a free electron. The rest energy of the electron-positron pair is released in the annihilation and included in the table.

We learn from tab. 3.1 that the time scale for the first step is longer than for the other two steps by orders of magnitude. This means that the first step is the bottleneck of the whole process, and any deuterium atom produced in this step proceeds immediately through the other steps. The reaction time of the whole process is that of the first step, and ^{2}D and ^{3}He exist in equlibrium.

We also observe that in the third step, two ^{3}He atoms participate. This means that the first and the second steps have to occur twice before a third step can occur once. Summing up the energies of the whole process (keeping in mind that the energies of the first and second steps should be taken twice), we find that the p-p process releases a total amount of 26.729 MeV for each helium atom produced in the process. The first step releases a neutrino. This particle is electrically neutral, with negligible mass (probably zero), whose cross-section for interaction with matter is very small. Actually, a star of the order of the Sun is transparent to this particle. It leaves the star with energy which does not convert to thermal energy. Its energy should therefore be subtracted from the energy contribution of the process. The neutrino produced in the first step of p-p reaction leaves with an average energy of 0.26 MeV. This energy (times two) is subtracted to yield 26.2 MeV for the p-p process.

The process described in tab. 3.1 is one variation of the p-p process, called PPI. There are two parallel channels, named PPII and PPIII, where instead of the third step in which two ^{3}He atoms interact to form ^{4}He (and two ^{1}H), the ^{3}He interacts with ^{4}He or ^{1}H to form ^{4}He through intermediate steps involving beryllium and boron. The important difference between the three channels is that in PPII a neutrino with energy of 0.8 MeV leaves the system, whereas in PPIII a neutrino with 7.2 MeV does the same. Thus these two channels contribute less energy to the energy balance in the star. The relative weight of PPII and PPIII increases with the temperature, but at the same time the total weight of the p-p channel decreases relative to the carbon cycle process (see below), which becomes dominant at higher temperatures.

The exact calculation of the relative weight of the PPII and the PPIII channels is especially important for the measurements of solar neutrino flux because the higher energy neutrinos released in these subchannels are more accessible to observation on Earth. Observing solar neutrinos is important to us because they do not interact with the solar material and their properties directly reflect the conditions at the centre of the Sun. Observing these particles is actually seeing directly into the solar centre. The photons observed in solar luminosity, however, are produced at the surface and carry no characteristic signals about the conditions existing at the solar centre, the location of energy production.

A useful formula for calculating the energy balance in stars is that for the rate of energy production by p-p process, ϵ_{PP},[1] in units of erg gm^{-1} sec^{-1}:

$$\epsilon_{PP} = 2.32 \times 10^6 \frac{\rho X^2}{T_6^{2/3}} e^{-\left(\frac{33.81}{T_6^{1/3}}\right)} \left(1 + 0.0123 T_6^{1/3} + 0.0109 T_6^{2/3} + 0.00095 T_6\right)$$

(3.23)

where T_6 is the temperature in units of 10^6 degrees.

3.9 The Carbon Cycle (CNO)

In this reaction a carbon nucleus interacts with a proton to form an ^{13}N atom, which is an unstable particle and disintegrates immediately to form ^{13}C. The ^{13}C interacts with another proton to form an ^{14}N atom. This atom interacts with another proton to form ^{15}O, which disintegrates immediately to form an ^{15}N. This atom, in turn, interacts with yet another proton to form ^{12}C+^{4}He. In the overall process, a carbon atom is returned to the system while four protons are absorbed in the cycle to produce a helium atom. The carbon has the role of a catalyst without which the process cannot proceed. The process is summed in tab. 3.2.

Summing up the energy released in the process, we find: $\epsilon = 26.734$ MeV,

Table 3.2.

		energy(MeV)	τ(years)
$^{12}C + {}^1H$	$\rightarrow$ $^{13}N + \gamma$	1.954	8.92×10^5
^{13}N	$\rightarrow$ $^{13}C + e^+ + \nu$	2.221	2.76×10^{-5}
$^{13}C + {}^1H$	$\rightarrow$ $^{14}N + \gamma$	7.550	2.23×10^5
$^{14}N + {}^1H$	$\rightarrow$ $^{15}O + \gamma$	7.293	1.82×10^8
^{15}O	$\rightarrow$ $^{15}N + e^+ + \nu$	2.761	5.65×10^{-6}
$^{15}N + {}^1H$	$\rightarrow$ $^{12}C + {}^4He$	4.965	7.94×10^3

which is approximately equal to the energy released in the p-p process. Two neutrinos are produced in this process; one leaves with energy of 0.710 MeV, and the other leaves with energy of 1.00 MeV. These energies should be subtracted from the energy contribution of this process to the energy balance of the star.

There is a variation on this cycle, where in the last step $^{15}N + {}^1H \rightarrow {}^{16}O$ and proceeds to the production of ^{17}O. This atom interacts with a proton to produce a helium atom and an ^{14}N.

Whether the carbon cycle occurs depends on the presence of heavy elements — carbon and nitrogen. Thus first generation stars, which have a low percentage of metals, can barely maintain this process. In second and third generation stars, where the metals are a few percent of the stellar material, this process fulfills an important function. The rate of the carbon cycle depends much more on the temperature than the p-p process. Hence at low temperatures the p-p process is dominant. At a temperature of about 18 million degrees and two percent abundance of carbon and nitrogen, the rates of the processes are equal. At higher temperatures, the rate of the carbon cycle increases steeply and overcomes the p-p process.

The formula for calculating the rate of energy production by the carbon cycle, ϵ_{CNO}, is:

$$\epsilon_{CNO} = 8 \times 10^{27} \frac{\rho X \cdot X_{CN}}{T_6^{2/3}} e^{-\left(\frac{152.31}{T_6^{1/3}}\right)} \qquad (3.24)$$

where X_{CN} is the mass fraction of carbon and nitrogen. In fig. 3.3 the rate of energy production by PPI reactions and by the CNO cycle are compared as a function of the temperature.

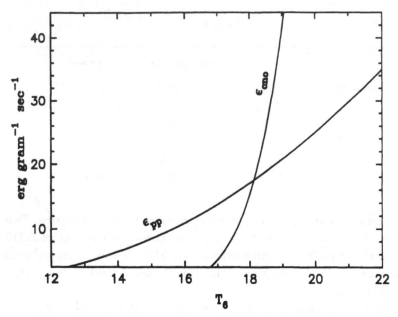

Figure 3.3. Rates of energy production by PPI channel and the CNO cycle vs. temperature, calculated according to eqs. 3.23 and 3.24. (T_6 is the temperature given in units of 10^6 degrees.)

3.10 The Helium Fusion

Upon the consumption of the hydrogen at the stellar centre, the star is left with a helium-rich core whose threshold for nuclear reactions is 80 to 90 million degrees. The stellar core, which has no nuclear energy source, contracts, and hydrogen continues to burn in a shell around the core. Helium burning starts when the temperature at the core reaches the threshold for this reaction. In an earlier section we discussed the case of degeneracy, for which the pressure depends on density alone and not on temperature. Helium cores of low mass stars (up to $2M_\odot$) develop degeneracy before reaching the temperature threshold for helium burning. This means that when burning starts, the first increase in temperature does not influence the pressure, the matter does not expand with increasing temperature, and the burning develops to a runaway (called a *helium flash*). Degeneracy is removed only when the temperature rises. The core then expands to lower the temperature to the equilibrium level of helium burning, at about 10^8 degrees. The peak temperature reached in the flash is about 10^9 degrees. This behaviour of flash ignition of a new stage in nucleosynthesis repeats itself in further nuclear burning stages. It is believed that medium mass stars explode when a flash occurs on the ignition of carbon-oxygen cores.

The helium burning is actually a fusion of three helium atoms to form a carbon atom and is called the *3α process*. The probability of the simultaneous collision of three helium atoms is very low. The process proceeds in two steps: in the first step two helium nuclei interact to form beryllium, and in the second step beryllium interacts with yet another helium nucleus to form carbon. The process is summarized in tab. 3.3.

Table 3.3.

		energy(MeV)
$^4\text{He} + {}^4\text{He} \rightarrow {}^8\text{Be} - \gamma$		−0.092
$^8\text{Be} + {}^4\text{He} \rightarrow {}^{12}\text{C} + \gamma$		7.366

The rest mass of ^{8}Be is higher than the rest mass of two helium atoms by an energy equivalent of 0.092 MeV. This is the amount of energy absorbed when beryllium forms from α particles, thus the negative sign of γ in the first line of the table. At the thermal conditions in the star, this atom is unstable and will spontaneously disintegrate to two helium atoms. Its lifetime as a beryllium atom is 2.6×10^{-16} sec. During this time interval it must collide with another helium atom to form the carbon atom. This last step releases 7.366 MeV. Together with the energy absorption of the first step, the energy yield of this process is 7.274 MeV per carbon atom produced.

The energy production rate of this process, $\epsilon_{3\alpha}$, is:

$$\epsilon_{3\alpha} = 3.9 \times 10^{17} \frac{\rho^2 Y^3}{T_6^3} e^{-\left(\frac{4294}{T_6}\right)}. \qquad (3.25)$$

It is easily observed that this expression is very sensitive to the temperature.

Comparing the energy production per nucleon in hydrogen fusion to that in helium fusion, we find that hydrogen fusion produces 10 times more energy per nucleon than does helium fusion. This means that at the same energy production rate, the time scale for the hydrogen burning phase is much longer than in all the other phases in stellar evolution. The main sequence phase, which is the phase of hydrogen burning at the stellar core, is about 90 percent of stellar life.

When the mass fraction of carbon increases as a result of helium fusion, another process of helium burning begins: the interaction of a carbon atom with helium to form an oxygen atom.

$$^{12}C + {}^4He \rightarrow {}^{16}O + \gamma. \tag{3.26}$$

The energy production of this process is 7.161 MeV per oxygen atom produced, and the rate of this process increases with the temperature. Thus at the first stage of helium burning, the main product is carbon. Later, with increase in temperature, oxygen forms at a higher rate. The product of helium burning is a carbon-oxygen core.

In principle, the nucleosynthesis might proceed further, up to the formation of iron, ^{56}Fe, which is the element with the lowest rest mass per nucleon. We shall not go into the details of this evolution. Low mass stars do not reach advanced stages in stellar evolution. For high mass stars, where the advanced stages may be relevant, this evolution takes place on a very short time scale. We shall be concerned only with the gross features of the advanced nuclear reactions phase.

References

1. Clayton D.D., 1968, *Principles of Stellar Evolution and Nucleosynthesis*, McGraw-Hill Book Company, New York.

Chapter 4

The Structure of a Star

4.1 Polytropes

Let us start by using eq. 2.7:

$$P'' + P' \left(\frac{2}{r} - \frac{\rho'}{\rho} \right) = -4\pi G \rho^2. \tag{4.1}$$

The advantage of this equation is that when we use the adiabatic equation of state it becomes an equation for one variable only, ρ. This treatment is historically important because Emden obtained the first concepts of stellar structure in this way in 1907. But its importance is not solely historical. Even today, when we want to get a first approximation of a stellar model, this method will yield reasonable results by the use of a suitable adiabatic exponent or of multizone models with different adiabatic exponents for each zone.

The adiabatic relation between the pressure and the density is:

$$P = K\rho^\gamma = K\rho^{\frac{1+n}{n}} \tag{4.2}$$

where K is a constant, γ is the adiabatic exponent, and n is called the politropic index. Gaseous spheres in hydrostatic equilibrium for which eq. 4.2 holds at each point are called *polytropes*. For convenience we furthern define $\rho = \lambda\phi^n$, where λ is a scaling parameter and the function ϕ is equal to unity at the centre. Thus λ equals the central density ($\lambda = \rho_c$). Using eq. 4.2 and the expression for pressure in an ideal gas at the classical limit, we can work out the adiabatic relation between density and temperature,

55

from which we note that ϕ is a temperature-like function. Equation 4.2 now reads:

$$P = K\lambda^{\frac{1+n}{n}}\phi^{n+1}. \tag{4.3}$$

Substituting this expression in eq. 4.1 yields a differential equation for ϕ:

$$(n+1)K\lambda^{\frac{1}{n}}\frac{1}{r^2}\frac{d}{dr}\left(r^2\frac{d\phi}{dr}\right) = -4\pi G\lambda\phi^n \tag{4.4a}$$

or in another form:

$$(n+1)K\lambda^{\frac{1-n}{n}}\left[\phi'' + \frac{2}{r}\phi'\right] = -4\pi G\phi^n. \tag{4.4b}$$

This expression is the Lane-Emden equation for the structure of a polytrope. The solution for a given value of index n is called the "Lane-Emden function of index n." The term $\frac{\phi'}{r}$ suggests that in order to avoid singularity at the origin, the first derivative of ϕ should vanish at this point. Equation 4.4 has analytic solutions for special values of n: $n = 0, 1, 5$. A more general solution can be obtained by numerical techniques, like the use of power series. Such a solution is found in the form of:

$$\phi = \sum_{m=0}^{N} C_m r^m. \tag{4.5}$$

To obtain this solution, eq. 4.5 is differentiated twice and the derivatives are substituted in eq. 4.4. Comparison of the coefficients of terms with the same powers of r yield the values of the coefficients in the power series. The demand that ϕ' vanishes at the origin means that only even powers of r should remain in the solution. For low values of n, the solution always decreases with increasing r. For $n < 5$, the function goes to zero at some value of r, $r = r_1$. It is clear that when $\phi \to 0$, the density and the pressure vanish too, and r_1, at which ϕ has the first zero, is the outer boundary of the configuration.

For a given value of n there are two constants which determine the solution: λ and K. K is determined by the adiabat on which the model lies, and λ is an arbitrary parameter which represents the central density.

What are the properties of the configuration?

The radius of the star is r_1. It can be written as $r_1 = a\xi_1$, where $a = \left[\frac{(n+1)K\lambda^{\frac{1-n}{n}}}{4\pi G}\right]^{1/2}$, and ξ_1 is a nondimensional number whose value is determined by the value of n. The values of ξ_1 as a function of n were calculated by Emden and are given in tab. 4.1.

Table 4.1.

n	0	1	1.5	2	3	4	5
ξ_1	2.4494	3.1416	3.6537	4.3529	6.8968	14.9716	∞

The value of $n < 5$ is an upper limit for a solution which still allows a value of ξ_1 at which ϕ goes to zero. The mass contained in the configuration can be found by integrating the equation for the mass continuity between the boundaries $\xi = 0$, to $\xi = \xi_1$:

$$M = \int_0^{r_1} dm = 4\pi \int_0^{r_1} \rho r^2 dr.$$

We find that the mass is given by:

$$M = -4\pi a^3 \lambda \xi_1^2 \phi'_{\xi_1} = -4\pi \frac{[(n+1)K]^{3/2}}{(4\pi G)^{3/2}} \lambda^{\frac{3-n}{2n}} \xi_1^2 \phi'_{\xi_1} \qquad (4.6)$$

where ϕ'_{ξ_1} is the value of the derivative of ϕ with respect to ξ at the point $\xi = \xi_1$. It is clear from eq. 4.6 that for $n = 3$, λ does not influence the mass.

The value of K depends on the relation between pressure and density, and it characterizes the specific entropy of the matter. The specific entropy does not change through adiabatic changes, and K remains constant. Using the expressions for the radius and for the mass given above, we can obtain interesting relations. Separating K from the expression for the radius and from the expression for the mass, and then multiplying these expressions, we find:

$$K = \frac{4\pi G}{(n+1)} \frac{1}{(4\pi \xi^5 \phi'_\xi)^{1/3}} M^{1/3} R \lambda^{\frac{2n-3}{3n}} = A M^{1/3} R \lambda^{\frac{2n-3}{3n}} \qquad (4.7)$$

where A is a constant whose value depends on the adiabatic polytropic index only. For $n = 3/2$, the exponent of λ vanishes, and we have $K \propto M^{1/3} R$. This expression means that while a star undergoes adiabatic changes, its radius varies in the direction opposite to the change in the mass. When a star loses mass at a high rate, its thermodynamic structure cannot follow the change in the mass. The evolution can be considered as an (almost) adiabatic one. In such a case, the loss of mass implies an increase in the radius. We shall return to this point when we discuss mass transfer in a close binary system.

Let us now consider a few cases of specific polytropes:
(1) An adiabatic convective polytrope.
(2) A radiative polytrope with radiation pressure.
(3) A fully degenerate polytrope.

An Adiabatic Convective Polytrope

Main sequence stars with masses below $0.4 M_\odot$ are believed to be fully convective stars. If the temperature gradient in these stars is very close to the adiabatic one, an adiabatic convective polytrope is suitable for representing such stars. In such a situation the radiation pressure is usually unimportant. The matter behaves like an ideal monoatomic gas, for which $\gamma = 5/3 \Rightarrow n = 3/2$. To estimate the stellar parameters, we can, for example, try a central density (λ) of 74 gm cm^{-3}, and a central pressure of 8.4×10^{16} dyne cm^{-2}. These values yield for K: $K = P/\rho^{5/3} = 6.44 \times 10^{13}$. Inserting these values for r_1 and M in the equations, we find: $r_1 = R = 2.47 \times 10^{10}$ cm $= 0.358 R_\odot$, and $M = 7.80 \times 10^{32}$ gm $= 0.394 M_\odot$.

A Radiative Polytrope with Radiation Pressure

We define β as the fraction of the pressure contributed by the gas pressure, P_g:

$$P_g = \beta P = \frac{k}{\mu m_p} \rho T. \qquad (4.8)$$

The radiation pressure P_{rad} is given by:

$$P_{rad} = (1 - \beta) P = \frac{1}{3} a T^4. \qquad (4.9)$$

From these two equations we find for the temperature:

$$T = \left[\frac{k}{\mu m_p} \frac{3}{a} \frac{(1 - \beta)}{\beta} \right]^{1/3} \rho^{1/3}. \qquad (4.10)$$

Thus for the total pressure, we have:

$$P = \frac{1}{\beta} \frac{k}{\mu m_p} \rho T = \left[\left(\frac{k}{\mu m_p} \right)^4 \frac{3}{a} \frac{(1 - \beta)}{\beta^4} \right]^{1/3} \rho^{4/3} = K \rho^\gamma. \qquad (4.11)$$

Assuming that β is constant throughout the star, we find for this case an adiabatic equation with $\gamma = 4/3 \rightarrow n = 3$. Such a model is called the *standard model*. Let us try to use this solution to describe a model of the Sun. We assume that the radiation pressure is about 0.005 of the total pressure. Substituting $\beta = 0.995$ in eq. 4.11, we find $K = 8.86 \times 10^{14}$.

For this case the mass does not depend on the central density. We now have for the radius: $r_1 = 1.38 \times 10^{10}$cm $= 0.2R_\odot$, and for the mass: $M = 6.76 \times 10^{33}$gm $= 3.4M_\odot$. If we try to relate the standard model to a nearly $6M_\odot$ stellar model, with a central density of 15 gm cm^{-3}, and $\beta = 0.98$, we find: $K = 1.384 \times 10^{15}$, $R = 2.27 \times 10^{11}$cm $= 3.24R_\odot$, and $M = 1.358 \times 10^{34}$gm $= 6.79M_\odot$.

To describe a realistic model of a star, we can use the first case to represent a fully convective model and the second case to represent a fully radiative model. Actual models of stars which normally contain both radiative regions and convective regions can be represented by a multizone model, where the radiative zones are described by the second case, and the convective zones by the first. Demanding continuity of the functions and their derivatives at the boundaries between the different zones is an important link in developing such models.

A Fully Degenerate Polytrope

We have seen in eqs. 3.11 and 3.12 that for a fully degenerate nonrelativistic gas $K = 0.991 \times 10^{13}$ and $\gamma = 5/3$. For the fully degenerate, fully relativistic case, $K = 1.231 \times 10^{15}$ and $\gamma = 4/3$. These relations can be used to calculate the properties of highly contracted objects such as white dwarfs. In such stars, the electrons become degenerate because the thermal pressure cannot support the star against contraction. If the mass of the star is high, a very steep pressure gradient is needed to counter the gravitational force, and the pressure and density reach very high values at the centre. We can assume that in the most extreme cases, after the star has reached the advanced stages in its contraction, the stellar matter becomes relativistic. Hence we shall calculate what value of the mass a fully relativistic, fully degenerate star can hold. For fully relativistic gas, $n = 3$; and the mass of the star does not depend on the central density. Substituting K and n in the expression for the mass, we find that in this case:

$$M = \frac{1.148 \times 10^{34}}{\mu_e^2} \text{ gm } = \frac{5.75}{\mu_e^2} M_\odot. \qquad (4.12)$$

In contracted stars no hydrogen usually remains, and the matter contains only helium or heavier elements. For these elements $\mu_e \simeq 2$. Substituting this value in eq. 4.12, we find:

$$M = \frac{5.75}{4} M_\odot \simeq 1.44M_\odot. \qquad (4.13)$$

This mass is called the *Chandrasekhar mass* after S. Chandrasekhar[1], who calculated it in 1939. This is a limit on the mass of configurations which do not contain a nonthermal energy source in their interior and must be supported by degenerate electron gas. In contracted stars with higher masses, the pressure created by degenerate electron gas cannot support the configuration. The star collapses then further to form a neutron star (or black hole).

Emden's solutions were useful in obtaining preliminary notions of stellar structure. They also furnished the knowledge needed to take the initial steps in understanding how stars are composed and to obtain realistic physical parameters of stars for comparison with observed phenomena. Later, when the physical processes governing stellar structure (such as the nuclear reaction rates and the physical processes that create the opacity) became clearer, astrophysicists used approximate solutions to the equations of stellar structure applying them to appropriate partial zones in a star. The demand for continuity of the functions and their derivatives at the boundaries between the different zones enabled a choice of more realistic solutions that might be suitable for describing the full structure of the star. A more detailed account of this method can be found in Schwarzschild's book.[2]

At the end of the 1960s and throughout the 1970s came the development of big computers with very large computing power. These machines could solve the equations of stellar structure numerically, with the values of the gas characteristics calculated at each point and fed into the computer codes. This quantitative change in computational power implied a qualitative change in the ability to calculate stellar structure. To date researchers have found highly detailed models and calculated complicated phenomena in stellar structure and stellar evolution in detail. We shall briefly describe here how such methods are used.

4.2 Numerical Methods

We have already seen that the equations of stellar structure are not a set of ordinary equations for a set of variables. Instead, they are coupled to each other, so that no individual variable can be solved without solving for all the others. In addition, the values of the gas characteristics required for substitution in the differential equations are themselves functions of the thermodynamic variables, the solutions of which we want to find. We overcome this problem by a process of iteration.

We make an initial guess as to the structure to serve as a starting model. We then calculate the gas characteristics at the thermodynamic conditions in the initial model. These values are fed into the equations, which are now solved to yield a second model. We believe this to be a better approximation to the "true" model than was the first hypothesized model. This second model now furnishes a new initial model for yet another stage in the calculations to yield a third model. If we are proceeding correctly each model will be closer to the true model. After a sufficient number of steps, we might find a model in which the differences between consecutive steps are small enough to assume that we have arrived at a stable solution. It is clear that making a "good" guess for the starting model is crucial to ensuring a reasonable calculation time for the convergence of the solution.

The numerical treatment of a set of differential equations can proceed in two main ways. One way is to integrate the equations numerically, starting at one boundary and integrating the equations continuously to the other boundary. The other way is called the *relaxation method* and treats the whole star entirely. For the first method we use the known boundary conditions as the initial conditions for the integration process.

In Chapter 2 we stated that there are four boundary conditions: the mass and the luminosity vanish at the centre, and the pressure and the temperature are practically zero on the stellar surface. The direction of integration — from the centre to the surface or from the surface inward — can be chosen arbitrarily, either for convenience, or according to a feature in the overall picture with which we happen to be concerned. We insert the known boundary conditions at the starting point of the integration as the initial conditions. For the two remaining conditions, which are unknown at this point, a guess is made and introduced into the equations. When the other boundary is reached by the integration, agreement with the values of the known boundary conditions is a measure of the accuracy of our first guess.

As an example, let us consider an integration procedure for a star with a given mass from the surface inward. We use the boundary conditions $P_0 = T_0 = 0$, and guess some values for the stellar luminosity and for the radius. When we integrate to the centre, we expect the mass and the luminosity to become zero. If these values are not reached, we keep repeating the integration again and again, using different guesses for the initial values of the radius and the luminosity, until there is satisfactory agreement with the boundary conditions. Better results can be obtained by integrating from both directions — from the surface inward and from the centre outward. The two integrations meet at some point between the two boundaries. The demand that they match continuously is a measure of their quality.

At each integration step, we calculate the gas characteristics by the appropriate formulae, using the values that were obtained in the preceding integration step for the thermodynamic variables. To improve the integration process, sophisticated integration methods can be employed, such as the use of intermediate values in the integration step or the predictor-corrector method. Detailed accounts of such methods can be found in any textbook on numerical analysis.

The main disadvantage of this approach is that it is numerically unstable. The changes in the gas characteristics in different situations occur very quickly, with small changes in the thermodynamic variables. Moreover, there are certain situations in which the gradients of the thermodynamic variables are very steep (as is the situation at the edge of the stellar core of a red giant). As a result, in these cases very small changes in the independent variable result in big changes in the dependent variables, and the solutions frequently diverge. In addition, at each time step many variables are calculated separately for the differential equation of a certain variable. These consist of the gas characteristics and other variables involved in the equation of that certain variable. When a solution diverges, it is difficult to analyze and find which one of the variables involved is responsible. It is almost impossible to determine how to improve the treatment of the problematic variable. Hence this method is used only to investigate the influence of a particular factor in stellar structure.

In general, when we search for a stellar configuration determined by the combined influence of all the variables, we use the relaxation method, which solves the equations for the whole star simultaneously. In this method we propose an initial complete solution for the whole star. By solving the differential equations for the whole star simultaneously, we find the necessary corrections for the star to "relax" to the "true" solution. The relaxation method is treated in detail in the section below.

4.3 The Relaxation Method

We take the stellar mass as the independent variable, instead of the radius. The mass is actually the initial intrinsic property of the star, while the radius depends on the thermodynamic state of this mass. We multiply eqs. 2.29 by $dr/dm = 1/4\pi r^2 \rho$ to obtain:

$$\frac{dP}{dm} = -\frac{Gm}{4\pi r^4} \tag{4.14.1}$$

$$\frac{dr}{dm} = \frac{1}{4\pi r^2 \rho} \qquad (4.14.2)$$

$$\frac{dL}{dm} = q \qquad (4.14.3)$$

$$\frac{dT}{dm} = -\frac{3\kappa L}{4ac(4\pi r^2)^2 T^3} \qquad \text{radiative zone} \qquad (4.14.4a)$$

$$\frac{dT}{dm} = \frac{(\gamma - 1)}{\gamma} \frac{T}{P} \frac{dP}{dm} \qquad \text{convective zone.} \qquad (4.14.4b)$$

Equation 4.14.2 turns out to be an equation for the radius as a function of the mass.

Using the spherical symmetry which we assumed to exist in the model, we divide the star into N concentric mass shells. The fineness of the division depends on the level of the details in which we are interested. We divide the star into unequal shell masses so that in regions of steep gradients of the variables we should have a finer division. In regions where the gradients are more moderate, a coarser division will suffice. The number of mass shells in a regular stellar model is usually about a hundred.

The differential equations should be satisfied at each mass shell. However, since the equations include gradients of the variables, the value of each variable at a given mass shell is coupled to the values of that variable at the shells adjacent to it. As is usual in numerical treatment of differential equations, finite differences replace the differentials. We designate the shells by a running index, starting from the centre and proceeding up to the Nth shell. The thermodynamic variables for each shell are defined at the centre of the shell mass; while variables such as the radius, the mass, and the luminosity are defined at the outer boundary of the mass shell. Thus r_i is the outer radius of the ith shell, and m_i is the mass enclosed in a sphere with radius r_i. The variables T_i and ρ_i are the temperature and the density at the centre of the ith mass shell.

To illustrate, let us treat eq. 4.14.1, which is the equation of hydrostatic equilibrium. We write $dP \rightarrow \Delta P$; $dm \rightarrow \Delta m$, and we define a quantity ϵ_i as the difference between the left-hand side and the right-hand side of the ith equation. For the point m_i, we have: $\Delta P = P_{i+1} - P_i$. The equation is written:

$$-r_i^4(P_{i+1} - P_i) - \Delta m_i \frac{Gm_i}{4\pi} = \epsilon_i \qquad (4.15)$$

where Δm_i is the mass difference between the centre of the $(i+1)^{th}$ shell and the centre of the ith shell. Note that we have N equations of this form. In a true solution for a stellar configuration in hydrostatic equilibrium, all the ϵ_is equal zero. Usually the first guess is not a true solution, and the ϵ_is represent the errors — or the degree to which the guessed solution deviates from the true one. The goal now is to find a way to use the error in eq. 4.15 in order to achieve the true solution. We look for corrections δr_is which when added to the radii will minimize the errors ϵ_is. We express ϵ_i as a function of δr_is, and thus eq. 4.15 results in equations for the corrections δr_is, needed to minimize the errors. There is an equation for each r_i of the N mass shells. But we observe that the equation for the ith shell includes P_{i+1}, so that each equation is coupled to the equation of the following shell.

The dependent variables which appear in eq. 4.15 are the pressure and the radius. These variables interconnect because if the radius of a certain mass shell changes then the density of that shell changes as well, owing to the change of the shell's volume. This change in the density implies a change in the pressure. (Note that the temperature is kept constant throughout the treatment of the equation of hydrostatic equilibrium, a point to which we shall return later.) The pressure of the ith shell, P_i, depends on r_{i-1} and on r_i because both these radii determine the volume of the ith shell. They therefore determine the density of this shell, ρ_i. In the same way P_{i+1} depends on r_i and on r_{i+1}. Thus if we add a correction δr_i to the radius r_i, we must consider the correction in the pressure caused by this change. We now write eq. 4.15 again, adding terms which include the corrections δr_i that must be added to r_i in order to balance the equation. The corrections for P_i are of the form:

$$\delta P_i = \frac{\partial P_i}{\partial r_{i-1}}\delta r_{i-1} + \frac{\partial P_i}{\partial r_i}\delta r_i = \frac{\partial P_i}{\partial \rho_i}\frac{\partial \rho_i}{\partial r_{i-1}}\delta r_{i-1} + \frac{\partial P_i}{\partial \rho_i}\frac{\partial \rho_i}{\partial r_i}\delta r_i. \quad (4.16)$$

Note that we obtain the derivative of the pressure with respect to the radius through the derivative of the density with respect to the radius: $\frac{dP}{dr} = \frac{\partial P}{\partial \rho}\frac{d\rho}{dr}$. In the derivative of the pressure at a certain shell, we should use the equation of state which is appropriate to the thermodynamic state of that shell. We should also check how to determine whether the equation of state at that location is that of an ideal gas, or of a degenerate state, and so on.

For the correction for r_i^4, we write: $(r_i + \delta r_i)^4 = r_i^4 + 4r_i^3 \delta r_i + \cdots$ ignoring terms with powers of δr_i. Equation 4.15 takes the form:

$$- (r_i^4 + 4r_i^3 \delta r_i)(P_{i+1} - P_i) - \frac{Gm_i \Delta m_i}{4\pi}$$

$$- r_i^4 \left(\frac{\partial P_{i+1}}{\partial r_{i+1}} \delta r_{i+1} + \frac{\partial P_{i+1}}{\partial r_i} \delta r_i \right) + r_i^4 \left(\frac{\partial P_i}{\partial r_i} \delta r_i + \frac{\partial P_i}{\partial r_{i-1}} \delta r_{i-1} \right) = 0.$$

$$(4.17)$$

We set the equation equal to zero because we have added correction terms including the corrections δr_is to the original equation, which does not equal zero.

We rearrange eq. 4.17, collecting terms including the δr_is, and separate the "zero order" part of the equation, $(-r_i^4[P_{i+1} - P_i] - Gm_i \Delta m_i / 4\pi)$, which is the left-hand side of eq. 4.15. We obtain the equation:

$$r_i^4 \frac{\partial P_i}{\partial r_{i-1}} \delta r_{i-1} - \left[4r_i^3 (P_{i+1} - P_i) + r_i^4 \left(\frac{\partial P_{i+1}}{\partial r_i} - \frac{\partial P_i}{\partial r_i} \right) \right] \delta r_i$$

$$- r_i^4 \frac{\partial P_{i+1}}{\partial r_{i+1}} \delta r_{i+1} = r_i^4 (P_{i+1} - P_i) + \frac{Gm_i \Delta m_i}{4\pi}. \qquad (4.18)$$

When the corrections vanish ($\epsilon_i = 0$), eq. 4.15 is satisfied with zero on its right-hand side. We use eq. 4.18 as an equation for the corrections. The corrections found through eq. 4.18 are added to the r_is to form the new r_is used to calculate the next step. It is clear that when the zero order part of the equation equals zero, all the δr_is vanish — no more corrections are needed and the hydrostatic equation is satisfied.

We can write eq. 4.18 in a compact form:

$$A_i \delta r_{i-1} + B_i \delta r_i + C_i \delta r_{i+1} = D_i \qquad (4.19)$$

where

$$A_i = r_i^4 \frac{\partial P_i}{\partial r_{i-1}}$$

$$B_i = -4r_i^3 (P_{i+1} - P_i) - r_i^4 \left(\frac{\partial P_{i+1}}{\partial r_i} - \frac{\partial P_i}{\partial r_i} \right)$$

$$C_i = -r_i^4 \frac{\partial P_{i+1}}{\partial r_{i+1}}$$

$$D_i = r_i^4 (P_{i+1} - P_i) + \frac{Gm_i \Delta m_i}{4\pi}$$

where D_i is the zero order equation of the ith shell.

We have here N equations for the N δr_is, but we observe that in each one of the ith equations three unknowns appear: δr_{i-1}, δr_i, δr_{i+1}. All the N equations should be solved simultanously. We write the N equations in the form of a matrix equation. An N-dimensional vector $(\delta \vec{R})$ for the unknowns δr_is, multiplied by an $N \times N$ matrix $(\mathbf{Q})$ of the coefficients A_i, B_i, C_i. This product yields the vector of the constant terms, $(\vec{D})$, which contains the D_is:

$$
\begin{pmatrix}
B_1 & C_1 & .. & .. & .. & .. \\
A_2 & B_2 & C_2 & .. & .. & .. \\
.. & A_3 & B_3 & C_3 & .. & .. \\
.. & .. & .. & .. & .. & .. \\
.. & .. & .. & .. & A_{N-1} & B_{N-1} & C_{N-1} \\
.. & .. & .. & .. & & A_N & B_N
\end{pmatrix}
\cdot
\begin{pmatrix}
\delta r_1 \\
\delta r_2 \\
.. \\
.. \\
\delta r_{N-1} \\
\delta r_N
\end{pmatrix}
=
\begin{pmatrix}
D_1 \\
D_2 \\
.. \\
.. \\
D_{N-1} \\
D_N
\end{pmatrix}.
$$

$$(4.20)$$

We note that A_1 and C_N are excluded from the matrix because they contain terms of index 0 and index $N+1$ which do not exist. We have here a tridiagonal matrix, which makes the treatment more simple. Formally eq. 4.20 can be written in a compact form:

$$\mathbf{Q}\,\delta \vec{R} = \vec{D} \qquad (4.21)$$

where $\mathbf{Q}$ is the coefficients matrix, $\delta \vec{R}$ is the unknowns vector, and $\vec{D}$ is the constant terms vector. From matrix algebra we know that we can multiply eq. 4.21 from the left by $\mathbf{Q}^{-1}$, which is the inverse matrix of $\mathbf{Q}$, to get:

$$\mathbf{Q}^{-1}\mathbf{Q} \cdot \delta \vec{R} = \delta \vec{R} = \mathbf{Q}^{-1}\vec{D}. \qquad (4.22)$$

Thus, finding the inverse matrix of $\mathbf{Q}$ and multiplying the vector $\vec{D}$ by this matrix yields the corrections vector $\delta \vec{R}$. This vector contains the corrections δr_is which are added to the r_is to approach the true solution. The new r_is are used in the zero order part of the equation for the next step in the iteration. When the zero order of the equation equals zero, the true solution is obtained and the δr_is will vanish. In real calculations we are satisfied when the largest of the δr_is is below a certain value chosen as the convergence criterion.

We mentioned earlier that by the procedure described above we find a satisfactory solution to the equation of hydrostatic equilibrium — the equation for the pressure gradient. During this process we keep the temperatures at each mass shell constant. As explained in Chapter 2, many different combinations of temperature and density profiles can construct a given pressure gradient. In order to obtain a unique solution for the run of the thermodynamic variables, we have to solve another equation which involves the temperature gradient.

An inspection of the temperature gradient equation, (eq. 4.14.4a), shows that even if the initial temperature gradient supplied the correct luminosity, the changes in the $r_i s$ and in the opacity (due to the changes in the density), will undoubtedly change the luminosity (or the demand on the temperature gradient). Thus, we must seek a new temperature run. To solve an equation which involves the temperature, we repeat the procedure used above. An appropriate equation for this purpose is the energy equation (eq. 4.14.3), which is the equation for the luminosity gradient.

We replace the differentials by finite differences: $dL_i \rightarrow \Delta L_i = L_i - L_{i-1}$; $dm_i \rightarrow \Delta m_i = m_i - m_{i-1}$. Equation 4.14.3 now reads:

$$L_i - L_{i-1} - q\Delta m_i = \epsilon_i. \tag{4.23}$$

Again we do not expect that the guessed solution will satisfy eq. 4.23 with $\epsilon_i = 0$. We add corrections using the dependence of luminosity on temperature to obtain the corrections we need to improve the solution by adding them to the zero order temperatures. We use the equation for the temperature gradient in the form given in eq. 2.17 for radiative transfer and in eq. 2.23 for convective transfer. The luminosity L_i on the surface of the ith shell is proportional to the temperature gradient at this point, which is likewise proportional to the difference between T_i and T_{i+1}. Thus we have for δL_i :

$$\delta L_i = \frac{\partial L_i}{\partial T_i}\delta T_i - \frac{\partial L_i}{\partial T_{i+1}}\delta T_{i+1} \; ; \; \delta L_{i-1} = \frac{\partial L_{i-1}}{\partial T_{i-1}}\delta T_{i-1} - \frac{\partial L_{i-1}}{\partial T_i}\delta T_i. \tag{4.24}$$

We find therefore that the corrections for eq. 4.23 for the ith shell include the corrections for T_{i-1}, T_i, and for T_{i+1}. The values of q and κ, which are the nuclear reaction rates and the opacities, are calculated using the values of the thermodynamic variables obtained in the preceding iteration. These values are inserted into the equations. Again we use this equation as the equation for the corrections, and we rearrange the N equations in the form of a tridiagonal matrix. The corrections found are added to the $T_i s$ which are to be used in the zero order terms in the next iteration. During this treatment we keep the radii (and the densities) constant, until a satisfactory solution is obtained for eq. 4.23.

Undoubtedly when the new values of the temperatures are inserted in the zero order part of the equation, the pressure will change from the value found by solving the hydrostatic equation. This equation, which was satisfied at the end of the iterations described in eqs. 4.15 to 4.22, is now disturbed. We return now to this equation and solve it again by the same method. However, we expect that the corrections needed now for the $r_i s$ will be smaller than those needed when we started the first iteration cycle. After solving the hydrostatic equation in the second cycle, the energy equation is disturbed, and we return for a second cycle of iterations to solve eq. 4.23.

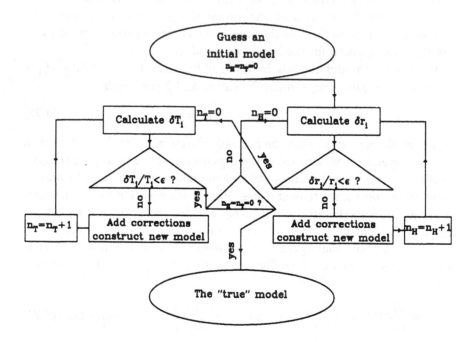

Figure 4.1. A flowchart of a computer code for computing stellar structure.

We thus solve the two equations alternately until we reach a configuration in which solving one equation does not disturb the solution of the other, and the corrections for both equations are below the convergence criterion. This configuration can be considered as the "true solution" for the structure of the star. The flowchart of the computer code summarizes the process of computation schematically in fig. 4.1. Note that we solved only two equations by iterations: the hydrostatic equation for the radii and the energy equation for the temperatures. Yet all the other equations were used. We used the equations of state and the equation of mass continuity for substitution in the hydrostatic equation. For substitution in the energy equation we used the equation of the temperature gradient, the equation of the nuclear reaction rates, and the equation of the opacity. Therefore, when finding the satisfactory solutions for the hydrostatic equation and for the energy equation, we satisfy all the other equations as well.

Let us summarize the relaxation method in brief. We solve two equations: the hydrostatic equation (the equation for the pressure gradient) and the energy equation (the equation for the luminosity gradient). Each equation is solved by a sequence of iterations until a satisfactory solution is found. When we solve the hydrostatic equation, which yields the new radii and densities, we keep the temperatures constant. After solving this equation we turn to the energy equation in order to solve for new temperatures, again by a sequence of iterations. Upon treating this equation, we keep the radii (and the densities) constant.

We treat the two equations alternately, until satisfactory solutions for both the radii and the temperatures are found. In each iteration we write the zero order equation, plus the terms which include corrections. This equation is used to find the needed corrections to improve the solution. We write the N equations for the N shells in the form of an $N \times N$ tridiagonal matrix. By inverting this matrix, we find the corrections for which we are searching. More details about this method may be found in the paper of Rakavy, Shaviv, and Zinamon.[3]

Clearly the amount of calculation needed for $N \sim 100$ is enormous. We use several cycles of iterations for each equation. Each cycle is composed of several iterations. In each iteration we have to write the equation for each of the N shells and substitute the values of the gas characteristics in the equations according to the local thermodynamic state.

No human being can hope to perform such a quantity of calculations in a lifetime. However, the method for performing these calculations is very routine. In each cycle of iterations, the steps of arranging the equations and calculating the values of the variables is the same. In each iteration, the calculations for each mass shell are performed identically and are done in the same way as in the one preceding it. In each iteration the calculations done for each mass shell are performed in the same way. Routine tasks of this nature are ideal for machine work; the large number of simple operations involved in such work suits the capacities of computers. Fast computers are now used to perform these calculations. The variables are arranged in vectors whose number of dimensions is the number of the mass shells. Each component of the vector contains the variable of the corresponding shell. For example, there is a vector for the temperature where each component in the vector contains the value of the temperature in the corresponding shell. Similarly, we have N-dimensional vectors for each of the variables. When we set out to perform an operation on all the shells, we arrange it in a loop form which repeats the calculations over the vector components. Any operation is performed in vector form over all the shells. Thus, the advantages of the machine overcome the problem of having to cope with the huge quantity of necessary calculations.

Today we have solutions for the configurations of many types of stars. We shall now describe some solutions for which the configurations were calculated by the relaxation method.

However a caveat is in order. From the recipe we described for the calculations, we might suppose that any initial guess can be used, and that the computer would do the work and find the true solution. Unfortunately this is not so. If the initial guess is too far from the true solution, the calculations may proceed in a path which takes us away from the true solution. The corrections needed would then increase from one iteration to the next. Here is one reason for such behaviour. When, for instance, we wrote $\delta L = \frac{\partial L}{\partial T}\delta T$, we used the first term in a Taylor series for the expansion of the function L around a certain point. The series contains terms with increasing powers of δT, with coefficients that contain increasing orders of the derivatives. We ignore the terms containing high powers of δT, assuming that $\delta T/T \ll 1$. An example of a similar treatment is the way in which we considered the correction for the term r_i^4 in eq. 4.14, where we wrote for the corrected r_i: $r_i^4 \rightarrow (r_i + \delta r_i)^4 = r_i^4 + 4r_i^3\delta r +....$, ignoring terms with powers of δr_is . This can be justified only if indeed $\delta r_i/r_i \ll 1$. Otherwise, terms with powers of δr_i cannot be ignored, and the calculations will be much more complicated.

A second reason is that at each iteration step we substitute the values of the gas characteristics calculated with values of the thermodynamic variables found in the preceding step in the equations. If the starting solution is far from the true solution, the values of the gas characteristics inserted into the equations will be very far from their values in the true solution. This situation may lead to solutions which move further away from the true solution.

A good guess for the initial model is therefore more efficient for carrying out the calculations and is sometimes crucial for the chance of finding any solution at all. Those who are well experienced in calculating stellar models usually make a good guess and are able to arrive at the true solution within reasonable computing time. A polytropic model constructed by the method described in Section 4.1 can serve as a good initial guess.

Another word of warning is worth mentioning here. We can obtain a "good" solution for a stellar structure where all the equations are satisfied, but the nature of this solution depends also on the energy content of the model. This energy content is actually introduced by the initial guess of the model. If the calculations for the true solution are performed correctly and maintain the conservation of energy, this content will not change. But this energy content was chosen arbitrarily, and different choices may be made. We can only arrive at a correct model with correct energy content by simulating the evolution of the model from a zero age model, including calculations of energy balance at each evolutionary step.

4.4 Stellar Models

Let us next consider a few characteristic models that can teach us about how different factors influence stellar configurations.

The Solar Model

We believe that our Sun has been in existence for about 4.5 billion years. We can calculate how much hydrogen has converted to helium during this period if we assume that solar luminosity did not change much during the lifetime of the Sun. The energy flux radiated by the Sun is 3.9×10^{33} erg $\sec^{-1}$. If each gram of hydrogen releases 6×10^{18} erg when it is converted to helium, then during the 1.5×10^{17} sec of solar life, 10^{32} gm of hydrogen were converted to helium. This quantity is about 0.05 of the solar mass. Nuclear burning takes place only at the hot solar centre, where the temperature is above the threshold for nuclear reactions. Most of the nuclear reactions take place in the core, which is about 0.15 of the solar mass. In the initial composition of the Sun, the hydrogen fraction was about 0.7. Thus in the initial core there was about 0.1 solar mass of hydrogen. Since 0.05 solar mass of hydrogen still remains in the core, there is enough nuclear fuel in the present phase for further burning at about the same rate for another 4.5 billion years. This indicates that the overall lifetime of the Sun, in the main sequence phase, is 9 billion years. When the hydrogen in the core is consumed, the Sun will switch to another phase in its evolution. We shall discuss this phase when we deal with red giant stars.

The temperature at the solar centre is 15.6 million degrees, and the

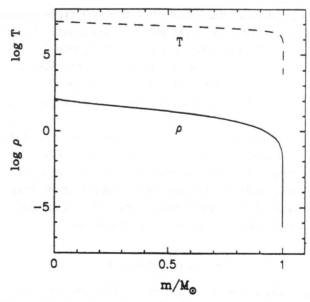

Figure 4.2. Density and temperature vs. mass in the present Sun.

density is about 148 gm cm^{-3}. The density, temperature, and pressure have negative gradients, which means that with increasing radius they decrease. In fig. 4.2 we display the run of these quantities as a function of the mass. The scale used for the thermodynamic variables is logarithmic, since their values run through several orders of magnitude.

From the temperature profile we learn about the locations of those processes which depend especially on the temperature: on the one end, we identify the high temperature region (above 12 million degrees) where the nuclear reaction rates are significant; and on the other end, we identify the regions of very low temperatures (below 10,000 degrees) where the matter becomes recombined.

From the density profile we learn that the central part of the Sun is a considerably dense region while the other parts are very dilute. The run of the mass vs. the radius shows that most of the solar volume is very dilute, and most of the mass is concentrated within a few percents of the volume.

Figure 4.3 displays the mass as a function of the radius for the present solar model. In the outer region of the Sun, the opacity increases with decreasing temperature until the temperature gradient needed to transfer the energy by radiation exceeds the adiabatic one (in absolute values), and

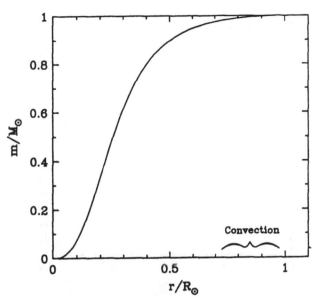

Figure 4.3. Mass vs. radius in the present Sun.

convection develops. The outer part of the Sun is convective, as is that of stars with masses of about one solar mass or less.

The nuclear reactions at the centre of the star are significant, although they actually start at lower temperatures of about four to five million degrees. The successive steps in the p-p reactions (tab. 3.1) depend on the temperature in different ways. The first step (^{1}H + ^{1}H $\rightarrow$ ^{2}D) begins at four to five million degrees. The second step (^{2}D + ^{1}H $\rightarrow$ ^{3}He) occurs immediately at a temperature of two million degrees. The third step (^{3}He + ^{3}He $\rightarrow$ ^{4}He + ^{1}H + ^{1}H) is much slower at low temperatures; below eight million degrees a great part of the burning hydrogen is converted to ^{3}He and does not proceed further to ^{4}He. Thus around the stellar centre, where ^{4}He is mainly produced, there is a periphery where the main product of hydrogen burning is ^{3}He. The fate of this material is an interesting question.

A Low Mass Star

We present an initial model of a star of $0.4 M_\odot$ at the very moment when nuclear reactions begin at the stellar centre. This stage is called the *Zero Age Main Sequence* (ZAMS). The radius of this star is $0.434 R_\odot$, the central

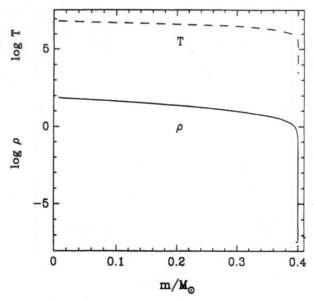

Figure 4.4. Profiles of the temperature and density vs. mass in a ZAMS star of $0.4M_\odot$.

density 73.4 gm cm^{-3}, and the central temperature 7.4 million degrees. The run of the thermodynamic variables vs. the mass is displayed in fig. 4.4.

We observe that owing to the lower central temperature, the nuclear energy yield is very low, and stellar luminosity is about $0.015L_\odot$. The lifetime of such a star in the main sequence phase is about 200 billion years, which is much longer than the lifetime of our Galaxy. This means that such a low mass star cannot complete its lifetime in the main sequence phase even if it is a first generation star. The temperature gradient needed for the energy transport is moderate. Although the central temperature is significantly lower than in a star of one solar mass, the region with a temperature of above four million degrees in such a star is relatively large. Thus the periphery around the core, in which ^{3}He is produced by nuclear reactions instead of ^{4}He, is significantly large and the product of the nuclear reactions is mostly ^{3}He material.

We mentioned earlier that in a low mass star convection occurs over a large area, and such a star can even be fully convective. The ^{3}He produced in the medial temperature zone is transferred by convection to the surface. If such a star loses mass from its surface at some stage in its evolution, the interstellar matter may be enriched by ^{3}He. This may be the cause for the abundance of ^{3}He observed in the interstellar matter, which is greater than the abundance expected in the primordial composition. However, no satisfactory theory or evolutionary calculations exist to date which can explain the present abundance of ^{3}He in the interstellar matter.

Comparing the last two models to the parallel models obtained using the polytrope scheme in Section 4.1, we find, as expected, that the model of the fully convective star agrees better with its parallel polytrope than does the solar model.

A Medium Mass Star

For the class of medium mass stars, let us describe a star of $6M_\odot$ at its ZAMS stage. Owing to high gravity, the pressure gradient needed to support the star is very steep, and the star is very hot at its centre. Its central temperature is 27 million degrees, and its central density is 15 gm cm^{-3}. Such a star is usually a second or a third generation star. We assume that its composition includes three percent of heavy elements, and the main nuclear reaction process is undoubtedly the carbon cycle. The luminosity of the star is $890L_\odot$, its radius $2.89R_\odot$, and its effective temperature 18,800 degrees. Most of the nuclear energy is produced at the very centre of the star. Because of the small surface area $(4\pi r^2)$ of the volume in which this energy is generated, a very steep temperature gradient is required if the energy is to be transferred by radiation. The temperature gradient in the stellar centre becomes steeper than the adiabatic one, and the core becomes convective. About 0.2 of the stellar mass $(1.2M_\odot)$ is in the convective core. Outside of this core the star is radiative up to the surface. With an effective temperature of 18,800 degrees, the star is fully ionized. The lifetime of such a star in the main sequence phase is about 56 million years, and it cannot be a first generation star. In fig. 4.5 we present the profiles of the temperature and density vs. mass in such a model.

This model compares quite well with the standard model of similar mass, calculated in Section 4.1.

We have described three models of main sequence stars — that of a low mass star $(0.4M_\odot)$, that of a low-medium mass star (the Sun), and that of a medium mass star. Let us turn now to a consideration of two stellar models at later phases than the main sequence phase: a red giant star and a white dwarf star.

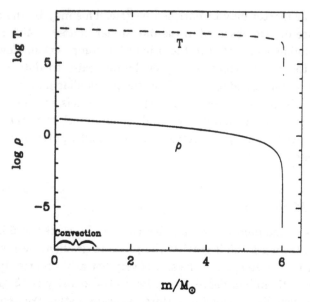

Figure 4.5. Density and Temperature vs. mass in a medium mass star.

4.5 The Red Giant

The model we present here is of a red giant with one solar mass, having a core mass of $0.58M_\odot$. We shall follow very briefly how a star reaches such a configuration. (A detailed explanation will be given later when we come to discuss processes in stellar evolution.)

The main sequence phase is the one in which hydrogen burns at the stellar centre. Upon the consumption of all hydrogen, the main sequence phase ends. The hydrogen now burns in a shell around a helium core. Since the number of particles per unit mass $(1/\mu)$ in heavy elements is lower than in hydrogen, density must increase to compensate for the decrease in pressure due to the decrease in $1/\mu$. The increase in density is reached by a contraction of the core. The contraction reduces the core radius. Therefore, the gravity at the core edge increases, demanding further increase in the pressure gradient. This demand leads to greater contraction of the core.

The process can continue unless degeneracy develops in the core, and the pressure gradient, created by a degenerate electron gas, supports the star against further collapse. The contraction of the core releases gravitational energy, part of which is absorbed in the envelope and causes the envelope to expand. Thus we find a configuration with a dense core and an expanded envelope. As a result of the expansion, the stellar surface cools down and the effective temperature decreases until most of the stellar radiation is at the red end of the spectrum. This is why such stars are red. Because of their expanded envelope, they are also giants — red giant stars. Owing to the continuous burning of hydrogen, the mass of the helium core increases with the hydrogen burning shell proceeding outward in mass. As the core increases, its temperature rises until it reaches the threshold for helium burning. Then helium starts to burn at the centre of the star. This phase is called the helium main sequence phase. Since the energy yield per unit mass in helium burning is about 10 times lower than that of hydrogen burning, and the stellar luminosity at the helium main sequence phase is higher than that of the hydrogen main sequence, the lifetime of a star in this phase is shorter than in the hydrogen main sequence phase by a factor of few hundreds. When the helium at the centre of the star is consumed, a contraction of the core and an expansion of the envelope again takes place, once more to form a red giant star. This phase is called the *asymptotic giant branch* (AGB), because its location in the H-R diagram is parallel to that of the red giant branch (RGB). The difference between the stars in the two branches is that in AGB stars there are two burning shells, an inner shell of burning helium and an outer one of burning hydrogen. The core is composed of carbon and oxygen (and heavier elements), and it increases gradually.

The star we shall now describe is an AGB star with two burning shells and a CO core. The core contains $0.58M_\odot$, and at the core's edge helium burns in a shell at a temperature of 120 million degrees. The mass between the two burning shells is about $0.01M_\odot$ and consists of helium rich matter. The stellar radius is $350R_\odot$, its luminosity $8,600L_\odot$, and its effective temperature 3,000 degrees. In fig. 4.6 we display the profiles of the density and temperature vs. the mass. The core edge is identified by the steep gradients of the two variables. These steep slopes are needed to create the steep pressure gradient demanded to support this configuration against collapse. It is interesting to note that the core, whose mass is about 60 percent of the stellar mass, has a radius of $0.1R_\odot$, which is less than one-thousandth of the stellar radius. We actually have two entirely different configurations: one is the dense core with densities of above 10^4 gm cm^{-3}; the other is the dilute envelope, whose density starts at 10^{-2} gm cm^{-3} close to the core edge and falls to 10^{-11} gm cm^{-3} near the stellar surface. Most of the stellar mass is within the core, while most of the stellar volume is taken up

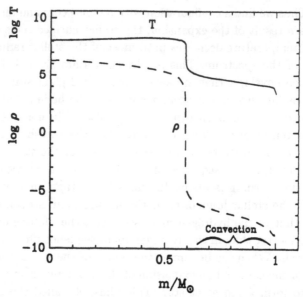

Figure 4.6. Density and temperature vs. mass in an AGB star.

by the envelope. All the processes involved in stellar evolution take place at the thin boundary layer between the two configurations.

An interesting feature is the temperature inversion at the central part of the star. Losses of energy from this region by neutrinos cause this inversion. The neutrinos are formed by interactions in the plasma which become significant at temperatures of above 70 million degrees. We mentioned in Chapter 3 that neutrinos hardly interact with matter, and stars are practically transparent to these particles. Hence they are radiated directly from the stellar centre where they are created, and this radiation cools down the central region. An inverted temperature gradient is formed, which causes a flow of heat from the burning shell towards the centre. Because of the inversion in the temperature gradient, the density gradient must become steeper so as to supply the pressure gradient needed to support the core against collapse. The matter in the core is highly degenerate due to the high density.

The envelope of the star, which is the region outside of the burning shells, is almost fully convective. This condition is due to the high opacities which exist in this region.

A special feature occurs toward the outer edge of the convective zone: a very steep temperature gradient and an inversion in the density. The reason for this interesting feature is the low efficiency of energy transport in this region. Because of the low temperatures, the opacity is very high, and radiative transport cannot be efficient. We observed in eq. 2.26 that the efficiency of the convective transport depends on the density and on the excess of the stellar temperature gradient over the adiabatic one (both in absolute values). As the density falls to very low values, a steep temperature gradient develops to compensate for the decline in efficiency. This is a "super-adiabatic" gradient. As a consequence of the steep gradient in temperature, the radiative transport is significant as well, and the radiative and the convective fluxes are of the same order of magnitude.

In addition to the demand on the temperature gradient, which is satisfied by a very steep gradient, there is a demand on the pressure gradient needed to support gravity. This demand is low since the gravity is low, due to the large radius. Hence the very steep temperature gradient creates an excessively steep pressure gradient, which is moderated by an inversion in the density gradient. We formulate this statement by using the expression for the pressure in an ideal gas:

$$P = \frac{k}{\mu m_p} \rho T \qquad (4.25)$$

$$\frac{dP}{dr} = \frac{k}{\mu m_p} \left(\rho \frac{dT}{dr} + T \frac{d\rho}{dr} \right). \qquad (4.26)$$

If the pressure gradient becomes too steep because of the temperature gradient, then a density gradient with an opposite sign will compensate for it. Thus we see how an excessively steep temperature gradient results in a density inversion. In fig. 4.7 we display the temperature and the density profiles in the outer part of the envelope.

From this region outward the star becomes radiative. A further decrease in the temperature causes a recombination of the plasma components to form neutral atoms. The interaction of radiation with neutral matter is negligible, and opacity falls by a few orders of magnitude. Radiative transport becomes very efficient. The temperature gradient becomes less steep than the adiabatic one, and convection vanishes. The overall picture of a red giant is of a star with a dense core which is inert with respect to nuclear reactions. All the energy production by nuclear reactions takes place in thin shells at the core's edge. The envelope which surrounds the core is very dilute and extends to a large radius with a low effective temperature on the surface. Most of the envelope is convective.

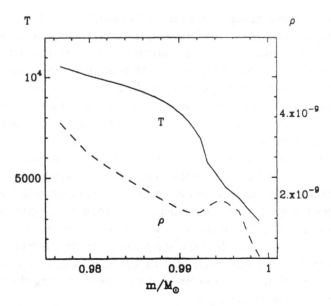

Figure 4.7. Density inversion in the outer part of the envelope of an AGB star.

4.6 The White Dwarf

The structure of a white dwarf star is very similar to that of the core of a red giant. We believe that a white dwarf is the remnant of a red giant which by some mechanism has lost its envelope. White dwarfs are composed of heavy elements (CO) or of helium, but do not contain hydrogen. The density at the centre of a white dwarf is around 10^5 gm cm^{-3}, and here no nuclear reactions take place. As it loses energy to the surrounding space, the star cools down. The cooling rate is proportional to the temperature of the cooling object. In the beginning, when the hot white dwarf is formed, its cooling rate is high. As it cools down, the rate of cooling decreases and becomes very low. A white dwarf star starts as a very hot star, with an effective temperature higher than 100,000 degrees, but cools down quickly to a few tens of thousands degrees. When the star is hot, its surface is bright (this is the origin of its name); but since its surface area is small, its total luminosity is low. When it cools gradually, its effective temperature decreases until it turns into a dark dwarf instead of a white one. At this phase the star cannot be observed. White dwarfs located in close binary systems may accrete hydrogen rich matter from their companion. When this matter ignites in a nuclear explosion, the star appears as a spectacular new entity, called a *nova*.

Most of the stellar population (over 90 percent) are main sequence stars. White dwarfs and red giants make up less than 10 percent of the stellar population.

References

1. Chandrasekhar S., 1939, in *Stellar Structure*, University of Chicago Press, Chicago.
2. Schwarzschild M., 1958, in *Structure and Evolution of Stars*, Princeton University Press, Princeton.
3. Rakavy G., Shaviv G., Zinamon Z., 1967, *Astrophys. J.*, **150**, 131.

Chapter 5

Computation of Stellar Evolution

In the preceding chapter, the relaxation method was described as a tool used in calculating stellar structure. When this method is used to simulate stellar evolution, some modifications are required.

5.1 Evolutionary Time Scale

We have seen that different processes of stellar evolution take place on three different time scales: dynamic, thermal, and nuclear. The dynamic time scale, where hydrostatic equilibrium is either reached or violated is of the order of a few thousand seconds. The thermal time scale, where thermal processes may change stellar structure, is of the order of a million years. The nuclear time scale, where the composition changes significantly due to nuclear reactions, is of the order of a billion years. (These values for the time scales are correct for low-medium stars. The thermal and the nuclear time scales are much longer in low mass stars, and much shorter in medium and high mass stars.)

We have to choose which time scale we shall use, making some assumptions about the other processes. It is convenient to select thermal evolution as the calculated process. The length of the evolutionary time step is chosen in such a way that the thermal variables change at each time step by only a small fraction (less than one percent). Thus we can expect that the simulated evolution will be close to the real evolution which is a continuous process. We assume that on this time scale, hydrostatic equilibrium is reached immediately at each time step, and the evolution is calculated as the model's movement through a sequence of (quasi) static states. The changes in the composition by nuclear transmutations are very small on the thermal time scale and are of the order of 10^{-5}.

How are the calculations performed?

Suppose we have a model in a hydrostatic and thermal equilibrium at some point along its evolutionary track. In such a model all the differential equations are satisfied. If there is no change in the composition, the model will remain static. However, such a stationary model includes some nuclear reactions that replenish the energy lost by radiation from the stellar surface. These nuclear reactions imply changes in composition, which in turn disturb the hydrostatic and thermal balance. Therefore, we should search for a new balanced configuration. There are two main reasons for the disturbances due to the change in the composition. One reason is the change in the pressure caused by the change in the number of particles per unit mass $(1/\mu)$. The other reason is the change in the heat capacity caused by the same change in $1/\mu$. The first directly influences the hydrostatic equation, while the second influences the energy equation. Thus we see that the nuclear transmutations in fact drive stellar evolution. The aging of a star is actually a change in its composition from light elements to heavier ones.

To calculate an evolutionary step, we start with a configuration in hydrostatic and thermal equilibrium in which the thermodynamic variables determine nuclear reaction rates. After determining the length of the time step, we calculate the changes in the composition by multiplying the rates of the nuclear transmutations by the time step δt. We insert the new compositions into the model, and calculate a new equilibrium state. Since we assume that the hydrostatic equilibrium is reached immediately, we solve the hydrostatic equation in the same way as for the static case (eq. 4.18). However the energy equation cannot be solved in the same way. We have to include the thermal processes such as the energy flow between mass shells, the changes in the gravitational potential energy, and the energy converted to work in expansion. Thus for the energy calculation we have to solve the full energy equation, which includes the time derivatives (eq. 2.27):

$$\frac{d\epsilon}{dt} = \rho q - \frac{1}{4\pi r^2}\frac{dL}{dr} - \frac{P}{V}\frac{dV}{dt}. \qquad (5.1)$$

Note that all the terms in this equation actually involve time derivatives, because q is the rate (time derivative) of the production of nuclear energy per unit mass, and $L(r)$ is the rate of energy flow from a surface of a sphere with radius r.

For the sake of convenience, we would usually make a transformation of variables here. Recalling that in order to find a unique solution for the stellar configuration, we must solve two equations for two thermodynamic

variables. In eq. 4.15 we solved the hydrostatic equation for the radii of
the mass shells, keeping constant the temperatures of these shells during
this treatment. (With constant masses of the shells, a solution for the radii
is actually a solution for the densities.) In eq. 4.23 we solved for the temperatures of the shells, keeping the radii (the densities) constant. A more
convenient choice is to use entropy as the second thermodynamic variable
instead of temperature. We do so because according to thermodynamic theory, when entropy does not change during thermodynamic processes, the
processes are adiabatic, and simple formulae connect the thermodynamic
variables. Thus if the entropy is kept constant while solving the hydrostatic
equation, the connections between the thermodynamic variables are calculated according to the adiabatic relations. Therefore we replace the energy
differential, $d\epsilon + \frac{P}{V}dV$, by the entropy differential multiplied by the density
and the temperature:

$$d\epsilon + \frac{P}{V}dV \Rightarrow \rho T dS \qquad (5.2)$$

where S is the entropy per unit mass. We add also a term to the energy
equation which includes the changes in the entropy due to the changes in
the composition. Multiplying the energy equation by $dV = 4\pi r^2 dr$ and
using finite differences instead of differentials, we obtain the equation:

$$T_i \Delta m_i \frac{\delta S_i}{\delta t} = q \Delta m_i - \Delta L_i + T_i \Delta m_i \sum_k \frac{\partial S}{\partial x_k} R_k \qquad (5.3)$$

where x_k is the abundance of the kth element in the composition, R_k is
the transmutation rate of the kth element, and the summation runs over
all the elements (the transmutation rates vanish for elements which do not
change). We keep the notation δ for time-dependent differences (such as
$dS/dt \Rightarrow \delta S/\delta t$) and Δ for mass-dependent differences (such as $dL/dm \Rightarrow$
$\Delta L/\Delta m$). The luminosity crossing the outer surface of the ith shell outward
depends on the temperature difference between the mass shells adjacent to
this surface. An increase in the temperature of the ith shell creates higher
luminosity across the ith surface, while an increase in the temperature of
the $(i+1)^{th}$ shell creates lower luminosity across the same surface. In the
solution of this equation, we are looking for corrections in the entropy. For
the corrections in the luminosity, we use the dependence of luminosity on
the entropy through its dependence on the temperature. We obtain:

$$\delta L_i = \frac{\partial L_i}{\partial T_i}\frac{\partial T_i}{\partial S_i}\delta S_i - \frac{\partial L_i}{\partial T_{i+1}}\frac{\partial T_{i+1}}{\partial S_{i+1}}\delta S_{i+1} = \frac{\partial L_i}{\partial S_i}\delta S_i - \frac{\partial L_i}{\partial S_{i+1}}\delta S_{i+1}. \quad (5.4)$$

We solve eq. 5.3 for the ith shell for corrections in δS_{i-1}, δS_i, and δS_{i+1}. Starting with a zero order configuration at time t, the resulting corrections are added to the zero order values of the entropies to form the zero order values of the entropies for the next iteration. We keep the densities (and the radii) constant throughout this treatment, until we obtain a satisfactory solution for the entropies whose corrections are below the convergence criterion. Then we treat the hydrostatic equation keeping the entropies constant. The two equations are treated alternately until satisfactory solutions are found for both the densities and the entropies. This solution is regarded as the configuration of the star at time $t + \delta t$. This configuration is used as the zero order model for the next time step.

At each step in the iterations the equations are arranged in a tridiagonal matrix as in eq. 4.20. They are solved by inverting the matrix and multiplying the vector of zero order variables by the inverted matrix.

Let us describe now the calculations for an evolutionary sequence. As already explained in the preceding section, the evolution is computed as the movement of the model through a sequence of quasistatic states. These states follow one another temporally, with a time interval δt between consecutive states. The age of the model at a certain state is the accumulation of the time intervals from the initial model (at zero age) to the particular state at hand. The model at each state differs from the preceding one by both a compositional change caused by the nuclear transmutations, and by the change in the star's energy content that took place during the last time interval. The changes in composition are calculated by multiplying the nuclear transmutation rates by δt. The energy changes are calculated by solving the energy equation (eq. 5.1 or eq. 5.3).

5.2 The Initial Model

Allowing a spherical gas cloud of given mass and composition to contract by its own gravitation can produce the initial model for an evolutionary sequence. This contraction causes an increase of the density and temperature of the cloud, until hydrogen burning at the centre of the cloud reaches the threshold temperature. At this stage the object achieves a steady-state equilibrium, and begins to function as a star in its main sequence phase. We consider this moment as the ZAMS and start counting the stellar age from this point.

The only initial conditions for a ZAMS star are its mass and composition which result from the mass and composition of the protostellar cloud. Thus we expect that a group of ZAMS stars formed with the same composition will differ from one another only by their masses. We must calculate the energy content of the star at its ZAMS stage which consists of the energy content of the primordial gas cloud forming the protostar, and the potential gravitational energy released in the cloud's contraction minus the radiated energy.

The internal energy content of the primordial cloud, E_c, is given by $E_c = \frac{M}{\mu m_p} kT$, where M is the cloud mass, $\frac{M}{\mu m_p}$ is the number of particles in the cloud, and kT is the thermal energy of each particle. For a composition of $X = 0.7$, $Y = 0.28$, and molecular hydrogen, we get $1/\mu \sim 0.42$. Assuming that the temperature of the cloud is 10 degrees Kelvin, we obtain as the energy content of one solar mass cloud: $E_c = 8 \times 10^{41}$ erg. The gravitational energy released in the contraction, E_G, is given by $E_G = \alpha G M^2 / R$ erg, where α is a fraction of the order of unity. Substituting the values of the solar radius and mass, we find $E_G = \alpha \, 3.8 \times 10^{48}$ erg. From the virial theorem we know that about half of this energy is stored as thermal energy in the star, and the rest is radiated. Thus the thermal energy content of the primordial cloud, E_c, can be ignored in the energy balance of the initial stellar model.

The composition of the star depends on the environment in which it was created. We believe that the primordial matter from which our Galaxy was formed contained mainly hydrogen and helium (with mass fraction for X, Y of about 0.7 and 0.3), and a negligible fraction of heavier elements (less than a mass fraction of 10^{-4}). First generation stars that were formed from primordial material contained a negligible percentage of metals. This has a bearing on nuclear processes in stars. Carbon is needed to catalyze the carbon cycle. In the absence or low abundance of carbon, even hot stars burn hydrogen via the p-p channel which is less efficient than carbon cycle burning. During stellar evolution the lighter elements transform to heavy elements by nuclear transmutations. When such stars lose matter to the interstellar medium either by continuous mass loss or explosion, the interstellar medium is enriched by metals. Stars formed later, in a second or a third generation, contain higher percentages of metals. Our Sun, for example, is considered to be a third generation star and contains about two percent metals.

5.3 The Boundary Conditions

The surface boundary conditions require special attention. In Chapter 2 we said that relative to the high values of the temperature and pressure in a star, the values of these variables on the surface are practically zero.

This is true, although considering the conditions for the radiation from the stellar surface requires a more accurate treatment. Photons formed by nuclear reactions at the stellar centre do not reach the surface in a "direct flight." On the way to the stellar surface, the energy is absorbed and re-emitted many times, usually in the form of lower energy photons. The average mean free path of a photon in the stellar interior is on the order of one centimeter. Close to the stellar surface, where density is very low and opacity falls drastically with the recombination of the atoms, the mean free path increases significantly. The outer shell, from which most stellar luminosity radiates outward without further absorption, is called the *photosphere*. To define the photosphere base at a radius R, we define the optical depth, τ, by: $d\tau = \kappa\rho dr$. Integrating this quantity from $r = R$ to infinity, the photosphere base is defined as the depth at which $\tau = 2/3$:

$$\int_R^\infty d\tau = \int_R^\infty \kappa\rho dr = \frac{2}{3}. \qquad (5.5)$$

In the Sun this depth is about 160 kilometres. The temperature at this point is called the *effective temperature* (T_e). The radius of this point (R) is defined as the stellar radius. The stellar luminosity is given by:

$$L = 4\pi R^2 \sigma T_e^4 \qquad (5.6)$$

where $\sigma = ac/4$ is the Stefan-Boltzmann constant. From these conditions we should obtain the boundary condition for the pressure. The hydrostatic equation at this point is:

$$dP = -G\frac{M}{R^2}\rho dr = g\rho dr \qquad (5.7)$$

where g is the gravitational acceleration. By multiplying and dividing this equation by κ, and assuming that the mass, radius, and opacity do not change much over this region, we obtain:

$$P_e = \int_\infty^R dP = \frac{g}{\kappa}\int_\infty^R \kappa\rho dr = \frac{g}{\kappa}\int_\infty^R d\tau = -\frac{2}{3}\frac{g}{\kappa} = \frac{2}{3}\frac{GM}{\kappa R^2} \qquad (5.8)$$

where P_e is the pressure at the photosphere base. Through this procedure we obtain the boundary conditions for the pressure and the temperature.

This method for calculating the boundary conditions of the star is quite simple and is suitable for calculating stellar atmospheres needed for models used in computing stellar evolution. When we want to calculate an exact atmosphere model that fits observations of an actual star, we need more sophisticated methods in which the contributions to the opacity in different wavelengths are calculated exactly.

5.4 The Evolutionary Code

The computation of stellar evolution consists of the solution of the two equations for the densities and for the entropies (4.15 and 5.3), and many auxiliary calculations performed in order to insert the values of the other variables and the gas characteristics for each mass shell at each iteration. The computer code should be constructed in such a way that the main track of the evolution is clear.

Usually a computer code includes a main subroutine and auxiliary ones. The main subroutine comprises the core of the calculations, which is the solution to the main equation in the matrix form. The auxiliary calculations are performed in different subroutines which the main subroutine calls when they are needed. At each iteration, the values of the zero order variables are substituted into the vectors which include the values of the variables at each mass shell. A subroutine to achieve this goal runs through all the shells. At each shell it calculates the new main variables by adding the corrections found in the preceding iteration. It then determines the secondary variables by calling special subroutines written for their calculation. For example, when calculating the pressure, the program must first determine which equation of state should be used in accordance with the local density and temperature — that is, whether this is to be the equation for an ideal gas in the classical limit or for a degenerate state, or a composite calculation carried out for a semidegenerate state, etc. According to this finding, the program summons the appropriate subroutine which calculates the pressure by using the local thermodynamic variables. It calls an appropriate subroutine in order to calculate the luminosity. This subroutine in turn calls another which calculates the local opacities at the shell that are needed for the computation of the luminosity.

The full set of variables is thus calculated for each mass shell at each iteration. The values of the calculated variables are substituted in the equations for the next iteration.

A crucial point in the calculation is the length of the time step. As explained earlier, the accuracy of the calculations in this method depends on the following assumption: the relative corrections to the variables are sufficiently small so that ignoring terms with higher orders of corrections will not introduce significant errors. The magnitude of changes in the composition and in energy (or entropy) content is proportional to the length of the time step. This length cannot be kept constant because the rate of change for the variables depends on local conditions in each shell and the changes in different evolutionary phases. The rate of compositional change and energy is continuously monitored. When these rates increase for any reason, we decrease the time step so as to avoid excessively large changes in the stellar structure in one step. When the rate of changes decreases, the time step can be increased to avoid unnecessary waste of computing time.

The division between mass shells needs careful monitoring as well. A large number of mass shells increases the computing time. However we want to approximate the continuous profiles of the variables in the star by a sufficiently fine division. We do not divide the star into equal mass shells. In a region where the gradients of the variables are steep, we have a fine division. In a region of moderate gradients, a coarser division may be sufficient. But this division is not kept constant. The fineness of the division should follow the steep gradient zones. Recall the profiles of density and temperature in a red giant star. These were extremely steep at the edge of the core, requiring a very fine division. But this feature does not remain at the same point in the mass. With the burning of light elements, the burning shells advance outward, and the core's edge advances with them. The region of fine division should follow this motion. A special subroutine checks, at every given number of time steps, the differences between the variables in neighbouring shells. If the relative difference between the variables in two neighbouring shells is larger than a chosen value, another shell is added between them. If, on the contrary, the relative differences are smaller than another chosen value, the two shells are united. If we follow an evolutionary calculation of a red giant star, we can observe how this subroutine divides and adds shells before the advancing burning front and unifies the shells behind this front. Special care should be given to maintaining energy conservation in such operations.

5.5 Dynamic Evolution

Up to now we have been dealing with evolutionary calculations for a quasi-static evolution which is simulated as a motion of the model along a sequence of hydrostatic configurations. The justification for such an approach is the ratio of the thermal time scale to the dynamic time scale, which is by some orders of magnitude greater for ordinary stellar evolution. But there are stages in stellar evolution in which this approach is unsuitable for describing the evolutionary track. In stages of an explosive nature, such as a nova explosion or helium flash in a red giant core, the changes occur very swiftly. Rapid alterations in the configuration abruptly alter the thermodynamic state in the star, so that the evolution can in no way be considered quasistatic. In such situations, we should calculate a fully dynamic evolution and solve eq. 2.1 without ignoring the term which contains acceleration. Dividing eq. 2.1 by ρ and substituting $1/\rho dr = 4\pi r^2/dm$ and $\ddot{r} = dv/dt$, we obtain the equation:

$$\frac{dv}{dt} = -4\pi r^2 \frac{dP}{dm} - \frac{Gm}{r^2}.$$ (5.9)

The left-hand side of eq. 5.9 is replaced by the difference expression for the ith shell: $dv/dt \Rightarrow (v_i - v_{i0})/\delta t$, where v_i is the velocity of the ith shell surface at the end of the time step δt, and v_{i0} is the velocity of the same shell surface at the beginning of the time step. v_i is defined by:

$$v_i = \frac{(r_i - r_{i0})}{\delta t} = \frac{\delta r_i}{\delta t}$$ (5.10)

where r_i and r_{i0} have the same relation to the time step as v_i and v_{i0}.

The finite difference form of eq. 5.9 is:

$$\frac{(v_i - v_{i0})}{\delta t} = -4\pi r_i^2 \frac{(P_{i+1} - P_i)}{\Delta m_i} - \frac{Gm_i}{r_i^2}$$ (5.11)

where Δm_i is the mass difference between the centre of the $(i+1)^{th}$ shell and the centre of the ith shell. Using eq. 5.10 to substitute for v_i, we find:

$$\frac{\delta r_i}{\delta t} - v_{i0} = -\left(\frac{4\pi r_i^2(P_{i+1} - P_i)}{\Delta m_i} + \frac{Gm_i}{r_i^2}\right)\delta t.$$ (5.12)

To this equation we add the expressions which include the corrections for the δr_is, as they appear in the left-hand side of eq. 4.17. Together with the δr_i, which already appear in eq. 5.12, we have in the equation for the ith shell the corrections for δr_{i-1}, δr_i, and δr_{i+1}. Again we arrange the N equations in a tridiagonal matrix and solve for the δr_is by inverting the matrix. By iteration we find the new r_is which will satisfy the equation. However now it is not a hydrostatic equation but an equation which also includes an acceleration term. The changes calculated for the radii are also created by velocities inferred from the accelerations.

This approach allows for fast changes in the configuration occuring at certain stages of the evolution where quasistatic calculation of the evolution is unsatisfactory. During such stages there are large changes in the energy content of the model. Huge amounts of energy are absorbed in a fast expansion or explosion of the star, or they are released upon collapse. Mass lost from the star during such stages may leave with energy in the form of kinetic or thermal energy of matter. Bursts of radiation can carry large amounts of energy. Usually, after such an event, the new configuration of the star is very different from the initial one.

The calculations of such stages have to be very precise and delicate. We do not know in advance what form the changes will take, nor what new configuration we can expect. The rapid changes taking place at such stages are very sensitive to the physical parameters and their evolution. Any inaccuracy in the calculations may therefore lead to a divergence of the solution. The choice of the convergence criterion can lead to rounding-off of errors that may divert the solution markedly. Calculations of such evolutionary stages require that special care be given to precision and place strict demands on the convergence criterion.

We can now see the considerable advantage of evolutionary calculations carried out by the quasistatic method. By employing this method, we looked for states of equilibrium. When such states exist, they are also states of local minima for the system's energy (which is why they are equilibria states). If we are not too far from the true solution, a correct calculation will lead us to it. If our solution diverges, it is usually because we tried a time step that was too long, and caused large changes which took us away from the local minimum in the energy. In such cases we can repeat the calculations using a shorter time step, and this will solve the problem in most instances.

5.6 Stellar Pulsations

In Chapter 1 we mentioned the regular pulsations of the Cepheids and their use in determining the distances to stars. The theory behind stellar pulsation is not altogether simple and straightforward. We shall present here only the outline of the theory.

The first approach to stellar pulsations was to treat them as a simple harmonic oscillator phenomenon in which conservation of energy keeps the system pulsating with a constant amplitude around the state of equilibrium. Such an approach assumes no loss of mechanical energy. However, no such system exists in nature. In any actual pulsating system the mechanical energy will eventually convert into heat and the pulsation will gradually decay. Astronomical systems, whose evolutionary time scale is of the order of millions of years, have to rely on driving mechanisms that will maintain the pulsations for a significant period of time.

We know from our preceding discussions that stability, which is actually a suppression of driving forces, results in part from the relation between pressure and temperature. The stabilizing mechanism can be understood as follows: heating brings about a pressure increase, and the pressure increase brings about expansion, which results in cooling. When the process begins with cooling, it initiates the opposite sequence. In a similar way the absorption of radiation brings about a stabilizing mechanism. In most cases the absorption coefficient is inversely proportional to some power of the temperature. Heating brings about a decrease in the absorption coefficient, which results in a better transfer of energy through the heated matter. This process likewise causes the matter to cool down. Cooling results in the same sequence in the opposite direction. A driving mechanism, designated by Eddington as the *valve mechanism*, is obtained by temporarily extinguishing the stabilizing mechanisms. Usually when matter contracts it loses energy to its environment by radiation, and when it expands it gains energy from its environment. The valve mechanism causes matter to gain energy when it contracts, and to lose energy on expansion. This peculiar behaviour is evidently limited to a certain range of the parameters of the system. When approaching the limit of this range, the "regular" behaviour of the matter is recovered.

Consider a certain region in the star that gains energy from ordinary stellar luminosity from its inner neighbouring region, and loses energy by a leakage to a region external to its location. If we stop the energy leak by closing an appropriate valve during contraction, the zone will not lose energy to its outer neighbour, and energy will accumulate. If we allow an increased leak by opening the valve during expansion, the zone will lose energy at an increased rate. The accumulation of energy at one stage and the increased energy leakage at the other furnish the driving mechanism for the pulsations. Let us now consider the valve mechanisms that can work in a star.

Two main mechanisms theoretically operate the valve: the γ mechanism and the κ mechanism. In the γ mechanism, the adiabatic exponent, γ, varies in a certain way that drives the pulsations. In the κ mechanism, the absorption coefficient per unit mass, κ, varies in a certain way that drives the pulsations.

For the sake of simplicity we treat only the radiative zones in which energy is transferred by radiation. This is called the *one zone model* treatment, suggested by Baker.[1] Here we discuss the behaviour of a confined zone assuming that the regions interior to this zone are unperturbed by the pulsations. Thus luminosity entering the zone from its inner neighbouring zone is constant and equals the luminosity of the equilibrium model, L_0.

A detailed mathematical treatment for the one zone model yields the equations from which these mechanisms can be derived.[1] We shall not go into the mathematical treatment. Instead, we shall try to sketch how these two mechanisms operate quantitatively.

The one zone model treatment assumes that the star exists in thermal and hydrostatic equilibrium. We study the behaviour of small perturbations around the equilibrium model. Assuming that only the small perturbations are time dependent, the equations including these perturbations are linearized to the first order in the perturbations. We assume that any perturbation, ξ, has a time dependency of the form:

$$\xi = \xi_0 e^{st} \tag{5.13}$$

where ξ represents the relative variations in the variable. Thus for a perturbation of the radius r : $\xi = \delta r / r$. From the linearized equations we obtain an algebraic equation of the third degree in s:

$$s^3 + K\sigma_0 A s^2 + \sigma_0^2 B s + K\sigma_0^3 D = 0. \tag{5.14}$$

If we consider only the outer part of the star, where no nuclear reactions take place, and the equation of state is that of an ideal gas at the classical limit, the constant coefficients in eq. 5.14 are given by:

$$\sigma_0^2 = \frac{Gm}{r_0^3}$$

$$K = \frac{2L_0}{\sigma_0(\gamma - 1) \cdot \Delta m \cdot c_V T} \tag{5.15}$$

$$A = (\gamma - 1)(4 - \kappa_T)$$

$$B = 3\gamma - 4$$

$$D = (\gamma - 1)[3\kappa_\rho + \kappa_T].$$

Variables with zero subscript have their values of the equilibrium model: Δm is the mass of the zone under consideration, c_V is the heat capacity at constant volume, and γ is the adiabatic exponent in the equation of state, and:

$$\kappa_\rho = \frac{\rho}{\kappa} \frac{\partial \kappa}{\partial \rho} \quad ; \quad \kappa_T = \frac{T}{\kappa} \frac{\partial \kappa}{\partial T} \tag{5.16}$$

where κ is the opacity per mass unit. The roots of eq. 5.14 determine the behaviour of the system. A third order equation has at least one real root. The two others are both either real or complex. If the real roots (or the real parts of the complex roots) are negative, then the perturbations will decay and the system is stable. No driving force develops in this case. The imaginary parts of the roots (if they exist) signify oscillatory behaviour. If the real part of the roots are positive, then the perturbations are expected to grow and a driving force develops for the pulsations.

We shall soon see that σ_0 is approximately the angular frequency of adiabatic pulsations of the one zone model. Hence by eq. 5.15 we find that K approximately shows the ratio of the energy flowing through the model to the internal energy of the zone during the time $1/\sigma_0$. A large value of K means that a significant fraction of the zone energy is lost during one period whereas low value of K means that a tiny fraction of the zone energy is lost during one period. In such a case, the process can be considered as almost adiabatic. K is a measure of the nonadiabaticity of the system.

To consider an adiabatic case, let us assume $K = 0$. With this assumption, the nonvanishing roots of eq. 5.14 are:

$$s = \mp\sigma_0\sqrt{-B}. \tag{5.17}$$

$B = 3\gamma - 4$, and it is clear that $\gamma > 4/3$ ($B > 0$) yields an imaginary solution for s, which provides oscillatory behaviour and a stable configuration where no driving mechanism develops. The pulsation period is $\sigma_0\sqrt{B} = \sigma_0\sqrt{(3\gamma - 4)}$. The condition for dynamic stability $\gamma > 4/3$ is a general one and is valid for any fluid structure. If $\gamma < 4/3$, then both roots of eq. 5.17 are real, and the system is dynamically unstable, and a driving force for pulsations exists.

For a nonadiabatic case, we consider $K \neq 0$. Two other conditions for stability are added to the one mentioned above:

$$\begin{aligned} D &> 0 \\ AB - D &> 0 \, . \end{aligned} \tag{5.18}$$

Consider first the condition $D > 0$. Assume that there is a "small" root $s_1 \ll 1$. Then the higher powers of s in eq. 5.14 can be ignored and we have:

$$s_1 = -\frac{K\sigma_0 D}{B}. \tag{5.19}$$

This root is a real number. If $D > 0$ (already assuming $B > 0$), the solution decays, retaining stability. The time scale for this decay is of the order of $\frac{1}{s_1} = \frac{c_V T \cdot \Delta m}{L_0}$, which is the thermal relaxation time (Kelvin time) of the zone.

From the second condition of eq. 5.18 we find the condition:

$$4(\gamma - 1) - \frac{4}{3} + [-\kappa_\rho - \kappa_T(\gamma - 1)] > 0. \tag{5.20}$$

Usually in an ideal gas at the classical limit, $\gamma = 5/3$. For most cases in the star, $\kappa_\rho \simeq 1$; $\kappa_T \simeq -7/2$ (see eqs. 3.15 and 3.16). Substituting these values into eq. 5.20, we find that $AB - D > 0$, and the stability condition is valid.

Let us consider when this condition may be violated. The first term in eq. 5.20 is always positive and contributes to stability. However this contribution is compensated by the second term, $-4/3$. The larger γ is, the greater its stabilizing contribution. In an ionization zone, γ approaches unity. Here the stabilizing contribution of the first term vanishes, and the negative term $-3/4$ prevails.

Writing the equation of state in the form:

$$P \propto \rho T \propto \rho^\gamma = \rho \cdot \rho^{\gamma-1} \qquad (5.21)$$

demonstrates that $\rho^{\gamma-1}$ represents the increase in pressure due to a temperature increase. In an ionization zone, $\gamma \to 1$ because the ionization process is endoergic. The matter in this zone absorbs energy without elevating the temperature. An ionization may absorb energy on contraction, while the pressure increase which should counter the contraction is delayed because the temperature does not increase. This mechanism is represented by the behaviour of γ ($\gamma \to 1$) and is hence called the γ *mechanism*.

The term in the square brackets represents the contribution of the absorption coefficient to stability and can be either positive or negative. Usually κ_ρ is positive and of the order of magnitude of unity; κ_T is usually negative. For $\kappa_\rho = 1$, $\kappa_T = -3.5$, $(\gamma - 1) = 2/3$, the square brackets yield $+4/3$, which is definitely a stabilizing value. With decreasing γ this term decreases and may even become negative, contributing to instability and pulsation driving. The influence of γ in this term is by means of the absorption coefficient, which causes radiation to be trapped in matter and thereby increases the energy content of the zone. The absolute values of κ_ρ and κ_T also decrease in the ionization zones, and κ_T may even become positive. This behaviour of κ contributes to the driving mechanism which is called in this case the κ *mechanism*.

The principal point that emerges is that the ionization zone plays an important role in defining the regime of pulsation driving. Owing to the absorption of energy in the ionization process, the matter gains energy during contraction without elevating the temperature. The consequence of releasing energy in the recombination process during expansion is that the matter radiates energy without lowering the temperature. The important issue now is the location of the ionization zone relative to the radiating surface of the star. A massive hot star, whose effective temperature is above 11,000 degrees, is entirely ionized, including the photosphere. In such a star no significant recombination can start, and it retains its stability against pulsation.

In a low mass cool star, whose effective temperature is below 7,000 degrees, the outer zones contain neutral matter, and the ionization front is well inside the star. It is kept in a steady equilibrium state between the ionization and recombination processes, which balance each other. Any valve mechanism which begins to operate in this zone due to random fluctuations is limited to the ionization zone. The neutral mass layers residing between this zone and the stellar surface operate as a buffer zone which prevents energy from leaving the star. This buffer zone acts in accordance with the regular thermodynamic laws which induce stability.

In a star whose effective temperature is in the range between the temperatures mentioned above, however, the ionization front is very close to the stellar surface. Suppose, then, that a small fluctuation at the stellar surface induces recombination in the ionization zone. The energy released in this process is radiated outward from the star, and the valve mechanism operates up to the stellar surface, driving the pulsation of the star. Thus we find that the candidates for pulsating stars are those whose effective temperatures are in the range of 7,000 to 11,000 degrees. The effective temperatures of the Cepheids are within this range.

The analysis of the pulsation mechanism described above depends on the assumption that energy is transferred by radiation. When convection takes place the considerations are different and more complicated. Convection which develops in a cool stellar envelope significantly enhances the efficiency of energy transfer. This is probably the cause of the stability of stars with effective temperatures of 5,000 to 7,000 degrees.

A special case is that of the long period variables called *Mira* stars, having effective temperatures of about 3,000 degrees and whose pulsational periods are of the order of years. These pulsations are possibly driven by the chaotic character of the system. A brief treatment of this topic will be found in Chapter 12.

References

1. Cox J.P., 1980, *Theory of Stellar Pulsations*, Princeton University Press, Princeton.

Chapter 6

Evolutionary Track

A solarlike star starts its existence with the contraction of a gas cloud as a result of gravitational attraction between its constituents. As we have seen in the preceding chapter, the initial energy content of the primordial cloud is very small compared to the gravitational potential energy released during the contraction. When the star contracts, the released gravitational energy heats the matter. The increase in temperature forms a temperature gradient, and the star begins radiating. When the protostellar material is neutral its opacity is very low, and the energy released can easily flow outward. This energy release causes a very rapid contraction which is almost like a free-fall motion. If we assume that the density in the primordial cloud is uniform, then we have for $m(r)$ (the mass enclosed in a sphere of radius r): $m(r) = \frac{4\pi}{3}r^3\rho$. The gravitational acceleration at a point r, $g(r)$, is given by:

$$g(r) = -\frac{Gm(r)}{r^2} = -\frac{4\pi}{3}G\rho r. \tag{6.1}$$

This equation means that the inward acceleration increases with the radius, and the acceleration in the outer zones is stronger than in the inner ones. With the increase in density, the number of collisions between the particles increases. Their velocities, originally directed toward the centre, now become random — in other words, thermal velocities. The temperature rises. With increasing temperature and density, pressure forms whose gradient slows down the contraction. According to the virial theorem, the kinetic energy of the particles in an ideal gas within the classical limit is minus half of the potential energy.

99

6.1 The Hayashi Track

During heating, the matter particles first dissociate from molecules into atoms, and the atoms are then ionized. These processes absorb great quantities of energy. Nevertheless the process of contraction is swift, and the luminosity of the star is higher, by one or two orders of magnitude, than in the main sequence phase. Since the stellar surface is cool, its radiation is mainly in the red end of the spectrum. At this stage the star is a red giant which starts with a large radius and high luminosity and contracts with a decrease in its luminosity. Hayashi[1] found that for each given mass and composition, the star cannot reach hydrostatic equilibrium at this stage if its surface temperature is below a certain value. The limiting values form an almost vertical line in the H-R diagram which is the right boundary for the evolutionary track of a star at this stage. When the protostar contracts and its luminosity decreases, it descends along this line called the *Hayashi track*.

The statement of Hayashi is based on calculating the structure of fully convective stars with a thin radiative surface layer. As we have seen in Chapter 4, a fully convective star can be described as a polytrope of index $n = 3/2$. For a star which is in hydrostatic equilibrium, the condition of equilibrium resulting from the hydrostatic equation determines the pressure gradient. Since the star is a polytrope, the profiles of all other thermodynamic variables are derived from the pressure gradient by the adiabatic relations. The only freedom remaining is in the value of the boundary conditions on the surface. The boundary condition for the pressure on the surface in a hydrostatic equilibrim is given in eq. 5.8:

$$P_e = -\frac{2}{3}\frac{g}{\kappa} = \frac{2}{3}\frac{GM}{\kappa R^2}. \tag{6.2}$$

Composition and temperature determine the opacity in a low density zone such as that found near the stellar surface. The pressure is related to the temperature by the appropriate equation of state. Thus for a given mass and composition which results in a given opacity, eq. 6.2 determines a relation between the stellar radius and the effective temperature. This relation yields the luminosity according to eq. 1.2. A detailed calculation yields the values of R and L for given mass and composition. The points in H-R diagram defined by these values form the Hayashi track. The conclusion is that a fully convective star with a radiative photosphere will be located on the Hayashi track.

What happens to a star that is accidentally located to the right of this track? This situation may occur if the stellar radius is greater than the value obtained from eqs. 6.1 and 6.2. Since eq. 6.2 was obtained by using the equation of hydrostatic equilibrium, its violation means disturbing that equilibrium. The star will contract to maintain the hydrostatic equilibrium, thus raising the effective temperature to the appropriate value. The area to the right of the Hayashi track in H-R diagram is called the "forbidden area" because no star in hydrostatic equilibrium can have its radius and effective temperature within the range defining this area.

The Hayashi track was calculated for fully convective stars. Stars which contain radiative zones will be located to the left of the Hayashi track. For fully convective stars, the average temperature gradient is about the same as adiabatic gradient. Stars to the right of the Hayashi track will have an average temperature gradient which is steeper than the adiabatic gradient. This will result in more rapid energy transport from the interior. The inner part of the star will cool down and the outer part will heat up, adjusting its temperature gradient in such a way that the effective temperature will maintain the appropriate value. If on the other hand the star is not fully convective, its average temperature gradient is more moderate than the adiabatic gradient, and the star will be located to the left of Hayashi track.

The overall scheme is as follows. As the star contracts it descends along the Hayashi track for as long as it is fully convective. Heating is faster at the stellar centre. With increasing temperature, the opacity at the central zone decreases. A radiative core develops which increases gradually with the heating, causing the star to move to the left of the Hayashi track. When the temperature increases above the threshold for hydrogen burning, a significant inner part of the star is radiative. The star then moves to its location on the main sequence.

The duration of the protostellar phase, from the beginning of contraction to the onset of the main sequence phase, is on the order of the thermal time scale, which for the Sun is about 30 million years.

6.2 The Main Sequence Phase

The main sequence phase is the period of hydrogen burning at the stellar centre. In the schematic figure for the evolution of the central density of a star (fig. 2.2), this phase is represented by an almost constant central density. This representation suggests that the thermodynamic conditions at the stellar centre do not change. However this is not exactly the case. The changes are very slow, occurring on a time scale of billions of years, but they are not negligible. We mentioned that the aging of a star is reflected by the transmutation of light elements to heavier ones. In the main sequence phase, this transmutation is of hydrogen to helium. This change results in a change in pressure due to the change in the number of particles per unit mass, $1/\mu$, which in ionized ideal gas is given by:

$$\frac{1}{\mu} = 2X + \frac{3}{4}Y + \frac{1}{2}Z. \tag{6.3}$$

Here X, Y, and Z are respectively the mass fractions of hydrogen, helium, and the metals. With the initial composition of the Sun where $X = 0.70$ and $Y = 0.28$, we have $1/\mu = 1.62$. When all the hydrogen is converted to helium, the same expression yields $1/\mu = 0.736$, which is less than half of the initial value. At the stellar centre, with a temperature at around 15 million degrees and a density of about 100 gm cm^{-3}, the gas is in a state of ideal gas, and the equation of state is:

$$P = \frac{k}{\mu m_p}\rho T. \tag{6.4}$$

A decrease of $1/\mu$ by a factor of two demands compensation by an increase in density and temperature. The changes are not linear because an increase in density increases the gravitational force acting on each mass element of the star. This in turn demands a steeper pressure gradient. Whatever the case, the star will adjust itself by increasing density and temperature, especially at the centre where the hydrogen is gradually consumed. These changes lead both to an increase in the nuclear reaction rates and to a steepening of the temperature gradient. The latter accelerates the energy flow outward. The reduction in the hydrogen fraction at the centre decelerates the nuclear reactions (eq. 3.10), but the increases in temperature and density overcome the deceleration. Thus the overall effect of the depletion of hydrogen is an increase in solar luminosity.

At the ZAMS stage, the luminosity of a solarlike star is $0.8L_\odot$. Toward the end of the main sequence phase, solar luminosity will probably reach the value of $1.4L_\odot$. This phase is represented in fig. 2.2 by a slightly inclined graph which reflects the slow increase in central density during the phase. Owing to these changes, there is a small change in the location of the star on H-R diagram: with the increase of luminosity and central density, the radius increases too. This means that the effective temperature will not change, and the star will move vertically upward in the diagram.

6.3 Post Main Sequence Evolution

When the hydrogen at the centre is entirely consumed, the changes in the stellar configuration become more significant. Because the temperature and density are greatest at the centre, the nuclear reaction rates in the solar interior are not uniform. At the end of the main sequence, we have a gradient of hydrogen fraction which starts from zero at the centre and increases gradually to the initial value at about a third of the solar mass. Nuclear burning at the centre is extinguished at this stage, and the hydrogen burns in a shell around the helium-rich core. This burning shell advances outward gradually, leaving behind the ashes of hydrogen burning — namely helium. Since no energy is produced in the core, it becomes almost isothermal. (If the core is not isothermal, the energy flow produced by the temperature differences will make it so.) A very shallow temperature gradient is still needed to transfer the gravitational potential energy released in the contraction. For a certain period this evolution proceeds steadily, with the hydrogen burning shell advancing outward, and the nearly isothermal core increasing accordingly. The density of the core increases as well. Owing to the flat temperature profile in the core, the density gradient becomes very steep and supplies the pressure gradient.

But this steady evolution terminates at a certain point. When the core reaches a mass of $0.14M_\odot$, it can hardly support its own gravity. Added to this gravity is the weight of the envelope which sits on the core's edge. In 1942 Schonberg and Chandrasekhar[2] showed that a core of $0.14M_\odot$ is at the limit of an isothermal configuration that can balance its own gravity. The additional weight of the envelope causes the core to collapse, and contract much faster than in a regular evolution. This collapse releases a great amount of gravitational energy and forms a temperature gradient to transfer this energy outward.[3]

The density increases with contraction, and this in turn increases the gravity, which accelerates the collapse. There is a positive feedback between the increase in density and the increase in gravitational force. Such a positive feedback is characteristic of a collapse process. This process stops when the high density creates a degenerate state for the electrons, and the pressure of the degenerate electron gas supports the core against further collapse. The core can continue to increase, supported by the degenerate gas pressure, up to the value of $1.44M_\odot$ — which is the Chandrasekhar mass (see eq. 4.13).

A part of the potential energy released in the contraction is deposited in

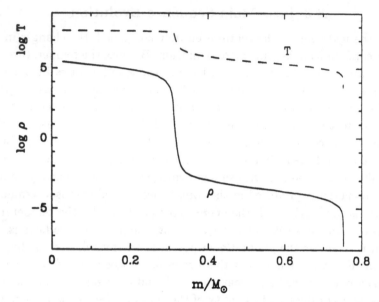

Figure 6.1. Density and temperature in a red giant star.

the envelope, and the envelope expands. With this expansion, the envelope cools down and the effective temperature decreases; the reddening of the stellar radiation reflects the latter. The opacities in the envelope increase and augment the demand on the temperature gradient until it becomes steeper than the adiabatic gradient. As a result, most of the envelope becomes convective. These changes cause the star to move to the right in the H-R diagram and approach the Hayashi track. Since its core is radiative, the star remains to the left of the Hayashi track. However, generally it repeats its course parallel to the Hayashi track, although it now does so in an upward direction. At this stage the evolutionary track of the star ascends in H-R diagram, parallel and close to the left of the Hayashi track. This evolutionary track is called the *red giant branch* (RGB). Along this track the helium-rich core increases and heats up. The envelope expands and stellar luminosity increases too.

6.4 Continuous Mass Loss

An important process which takes place during the RGB (and the AGB) phase is the continuous loss of mass from the stellar surface to the interstellar medium. This process was discovered through observation. Researchers found many red giant stars with significant surrounding gas clouds, which appeared, by the Doppler shift in their radiation, to be moving away from the stars with velocities of 5 to 30 km sec^{-1}.

From the density, ρ, of the cloud and its velocity, v, we can calculate the rate of the mass loss from the star, $\dot{m}$:

$$\dot{m} = \frac{dm}{dt} = 4\pi r^2 \rho \frac{dr}{dt} = 4\pi r^2 \rho v. \qquad (6.5)$$

If $\dot{m}$ and v are constant, we find an inverse relation between the cloud density and the square of its distance from the star r. With expansion the cloud becomes dilute until it fades into the interstellar medium. We shall return to this process later, so as to treat it in detail when we deal with planetary nebulae and related phenomena. For the present we shall consider only the importance of the mass loss process in stellar evolution. As it turns out, the masses of stars in their late evolutionary phases are lower than their initial masses. An important conclusion from this situation concerns the fate predicted for medium mass stars. Recall Chandrasekhar's limiting mass for white dwarfs (eq. 4.13), which is $1.44 M_\odot$. Higher mass stars, which evolve with constant mass, will end their evolution with white dwarf masses above this limit. These stars are doomed to collapse further, into denser configurations like neutron stars or black holes. With a mass loss whose rate increases significantly in later evolutionary phases, even stars with initial masses of $8 M_\odot$ may end their evolution as white dwarfs having masses below the Chandrasekhar limit. Indeed, for most observed white dwarfs, their masses are[4] within a narrow range of around $0.6 M_\odot$. From such an observation we conclude that some process exists which drives most stars to terminate their evolution with masses of approximately this value. The significant mass loss from red giants accords well with such a process.

It was found by Reimers[5] that the rate of mass loss from a star is proportional to the stellar radius and its luminosity, and inversely proportional to its mass. An empirical formula for the mass loss rate derived by Reimers from his observations is:

$$\dot{m} = -A\frac{LR}{M} \; M_\odot \; \text{yr}^{-1} \qquad (6.6)$$

where L, R, and M are the stellar luminosity, radius, and mass respectively. A is a numerical constant whose value should be determined to fit stars whose parameters are known. The minus sign shows that we are dealing with *loss* of mass from the star. A fit done for A by Reimers yields: for L, R, and M given in solar units, and $\dot{m}$ given in solar mass per year, $A = 4 \times 10^{-13}$.

From eq. 6.6 we observe that as the star climbs the RGB and its radius and luminosity increase, the mass loss rate also increases. The decrease in stellar mass due to the initial mass loss accelerates the mass loss further. The rate may increase by orders of magnitude along the RGB.

We are as yet unclear about the process which drives the mass loss. The outward acceleration of matter particles located a few stellar radii above the photosphere is reasonably well understood . The temperature above the photosphere of red giants is very low (only a few hundred Kelvin degrees). Particles of heavy elements stick together to form grains, so the metallic part of the matter exists in the form of dust. The nature of the grains depends on the composition of the metals. If the metals consist principally of carbon, then most of the dust will be made up of graphite particles. If the fraction of oxygen is equal to or greater than the fraction of carbon, then most of the dust will be CO particles.

However both types of grains are fully opaque, and photons which collide with such particles impart all their momentum to them (or twice the momentum, if the collision ends with a recoil of the photon). Thus the particles are driven by the radiation pressure. (Recall eq. 2.8, where we calculated the force that the radiation pressure exerts on matter.) The metal particles compose only a small fraction of the gas around the star. However, due to the coupling between the metal grains and the gas, the momentum imparted to the grains is shared with the gas, and the radiation pressure accelerates the whole volume outward.

The process described above is efficient in situations where most of the metals have already formed grains. This happens only at a distance of a few stellar radii from the stellar surface. It is not clear what kind of mechanism could drive the matter from the stellar surface such a great distance. There are number of proposals as to the nature of this mechanism, such as thermal instabilities in the envelope; shock waves at the photosphere base, or acoustic energy from the convection zone. None of these proposals has furnished a satisfactory explanation of the phenomenon. Stellar evolution is therefore computed by using the empirical formula of Reimers or similar empirical formulae. Research to discover the driving force behind the mass loss from the stellar surface continues.

The influence of the mass loss process on stellar evolution is very important. Low mass stars of below $0.75 M_\odot$ may lose their entire envelope while still in the RGB phase, before ignition of the helium at their centre. Such stars will turn into white dwarfs directly from the RGB phase and will become helium white dwarfs. Higher mass stars such as the Sun will lose about $0.2M_\odot$ while staying in the RGB phase. The helium ignites at their centre when they still have significantly hydrogen-rich envelopes.

6.5 Helium Ignition

When the temperature in the contracting core reaches the threshold for helium ignition at about 80 million degrees, the helium starts to burn. Nuclear reaction rates are low at the beginning, but the energy produced in these reactions heats the matter, and the nuclear reactions accelerate. At a temperature of 100 million degrees, the helium burning rate is quite significant. We mentioned earlier that in the case of low and low-medium mass red giants, there is a temperature inversion at their centre. The magnitude of the inversion depends on the mass of the star. The nuclear reaction rates depend on the temperature and density, but temperature dependence is steeper than density dependence. The helium burning begins close to the maximum in the temperature.

Starting with the initial contraction, and through all the evolutionary phases, higher mass stars are always hotter than lower mass stars in parallel evolutionary phases. We observe this situation when we compare the configurations of the three main sequence models described in Chapter 4. This observation has a very important consequence for the phase of helium ignition in the core of red giant: the temperature which develops in the core of high mass red giants is greater than in the core of low mass red giants. One result is that degeneracy develops to a considerably lesser degree in the core of high mass stars than of low mass stars. Red giants with masses equal or greater than $6M_\odot$ develop little, if any, degeneracy in their core.

A second important consequence is that since helium ignition depends on the temperature of the core interior, high mass stars reach their temperature threshold for helium ignition when the mass of their core is low. Low mass stars, however, have to build up larger cores before reaching this threshold. This circumstance is clearly displayed in fig. 6.2, which shows the core masses at helium ignition for several stellar masses. Since degeneracy becomes more marked with an increase of the core mass, evidence of this phenomenon is much higher in the cores of low mass red giants. The effect is somewhat obscured in medium mass stars because hydrogen burning in these stars during the main sequence took place in a convective core which mixed the composition to higher values of core mass. In a $6M_\odot$ star helium will ignite when the helium core is already $0.7M_\odot$.

The degeneracy in the core of low mass red giants causes the helium to ignite in a flash rather than burn steadily. We stated earlier that in a

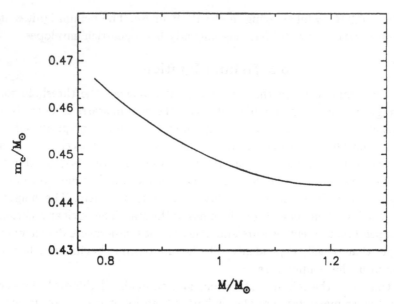

Figure 6.2. Core mass vs. stellar mass at helium ignition.

degenerate gas, pressure does not depend on temperature but rather on density alone. However the coupling between the pressure and the temperature is the thermostat, which stabilizes the star. This thermostat works only when the coupling exists. The thermostat mechanism may be described schematically as follows: when a system is in thermal and hydrostatic equilibrium, the sum of the forces at each point vanishes and the profiles of the thermodynamic variables are kept constant. If a nonthermal source somewhere within the system produces energy, the temperature gradients in the system transfer the energy away at the same rate it is produced. There is a negative feedback between the temperature and the cooling of the system such that, when some point heats randomly, the increased temperature creates higher pressure which causes immediate expansion. This expansion, in turn, cools down the heated point. If for some reason a local cooling takes place randomly, the pressure decreases and the resulting contraction heats the same place again.

The thermal equilibrium is not a static state, but it is maintained dynamically by the feedback mechanism between heating and expansion. The feedback is maintained throughout the coupling of temperature and pressure. As we saw in Chapter 2, the time scale for these dynamic adjustments over the dimensions of a main sequence star, or the core of a red giant, is a few thousands seconds. This time scale is very short compared to the thermal time scale, and we can think of it as an immediate adjustment. This immediate adjustment is also the thermostat which keeps the rate of nuclear reactions at exactly that of energy transport from the centre outward. If too much nuclear energy is produced, the matter heats up and the pressure increases. The matter then expands and cools down. If too little energy is produced, the reverse process takes place.

When the pressure is decoupled from the temperature, as happens in a degenerate state, the thermostat does not work. When helium starts burning in this state, the temperature rises which accelerates the nuclear reactions. The pressure does not rise, and no negative feedback acts to moderate the process. The positive feedback between the nuclear energy production rate and the temperature increase accelerates the process. Only when the temperature rises above 200 million degrees is the degeneracy removed, and the thermostat begins to work. But the nuclear reaction rates have already reached temperatures above the equilibrium point. They continue to increase before they slow down reaching temperatures above thousand millions of degrees, with nuclear energy production at a rate of 10^{11} solar luminosities. At this stage the core expands rapidly, and the cooling slows down the nuclear reaction rates. When the stellar core cools to about 150 million degrees, the helium burns in a steady-state equilibrium at the centre. With the expansion of the core, its edge cools down and extinguishes the hydrogen burning there. The envelope contracts, and the star maintains the *helium main sequence phase*.

The process, from the start of significant helium burning to the removal of the core degeneracy, is very rapid and has an explosive character. Huge amounts of energy are released during a very short time interval. What influence would such an event have on the outer part of the star? Several investigators carried out detailed evolutionary calculations so as to follow this evolutionary phase. They found that although the process at the central core has the characteristics of an explosion, this behaviour is limited to the inner part of the core. The effects of the explosion do not reach the envelope, which only "feels" the expansion of the core after the removal of the degeneracy. It responds to this expansion by a contraction and a decrease in the luminosity until the helium main sequence configuration forms.

The reason for this response is that the explosion-like event takes place in the central part of the core under the weight of a significant part of the core mass. This mass acts as a ballast which absorbs all the impressive effects emerging from the central part. The gravitational potential well of the core is so deep that the explosion at its bottom cannot lift the weight of the ballast. Thus we see no impressive effects in the star when the helium flash takes place.

6.6 The Horizontal Branch

When the helium burns at the stellar centre, the configuration of the star is similar to that of a star burning hydrogen at its centre. This phase is therefore called the *helium main sequence phase*. After hydrogen burning in the shell is extinguished, the envelope contracts and the effective temperature increases. The luminosity of the star is lower than that of the preceding red giant, but higher than that of a hydrogen main sequence star. Because of these changes the star moves to a new location in the H-R diagram. According to its luminosity the star is at the foot of the RGB and according to its effective temperature it is to the left of the RGB. The stars in this phase form a horizontal line which is the *horizontal branch* (HB) in the H-R diagram.

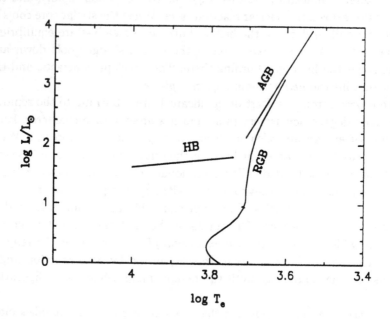

Figure 6.3. RGB, AGB, and HB in H-R diagram.

The exact location of the star on the HB depends on the remaining envelope mass. This mass depends on the initial mass and the amount of mass lost during the RGB phase. We mentioned earlier that in the RGB phase the star loses about $0.2M_\odot$ in a continuous process of mass loss. Its helium core is from 0.4 to 0.48 $M_\odot$, depending on the total stellar mass. The greater the stellar mass, the smaller the helium core is at the helium ignition. Stars with low mass envelopes have smaller radii and higher effective temperatures. They form the left end of the HB. Stars with high mass envelopes have larger radii and lower effective temperatures for the same luminosities. They form the red (right) end of the HB.

This phase continues as long as helium exists in the stellar centre. Because of the lower energy yield per gram from helium burning and the high luminosity of the star, this phase is much shorter than the hydrogen main sequence phase. When the helium at the centre is consumed, the star again turns into a red giant. However, it now has a CO core and two burning shells, one of helium and another of hydrogen. The evolutionary track in the H-R diagram is parallel to that of the RGB and is called the *asymptotic giant branch* (AGB).

6.7 AGB Evolution

During this stage the core further increases in mass, with stronger degeneracy. The star now has a larger radius and higher luminosity. Due to the energy losses from the centre by neutrinos, a temperature inversion again develops, covering a larger portion of the star. The maximum temperature is above 200 million degrees, and its location is close to the core edge. A deep temperature inversion exists toward the centre (see fig. 4.6). The luminosity increases significantly, as does the radius of the star. Thus the mass loss rate from the star becomes very high. The core increases due to the addition of the burning ashes, namely CO; and the rate of increase is proportional to the luminosity. The amount of energy released per mass in hydrogen burning is about 10 times greater than in helium burning. Therefore in calculating the core growth due to nuclear reactions, we consider mainly the production of energy by hydrogen burning. The rate of core growth, $\dot{m}_c$, is given by:

$$\dot{m}_c = \frac{\phi L}{X\epsilon}. \tag{6.7}$$

ϵ is the amount of energy released per unit mass of hydrogen converted to helium ($\epsilon = 6 \times 10^{18}$ erg gm^{-1}), X is the mass fraction of hydrogen in the envelope, and ϕ is a number which shows what fraction of the luminosity is produced by hydrogen burning. Equation 6.7 expresses the rate at which the envelope is consumed from its inner edge by nuclear reactions. From the outer edge, the envelope is consumed by the mass loss from the stellar surface. If we use Reimers' formula for the mass loss rate from the star (eq. 6.6), we have for the rate of decrease in the envelope mass, $\dot{m}_e$:

$$\dot{m}_e = \dot{m} - \dot{m}_c = -A\frac{LR}{M} - \frac{\phi L}{X\epsilon} = -L\left(\frac{AR}{M} + \frac{\phi}{X\epsilon}\right). \tag{6.8}$$

Both processes are proportional to the luminosity. Substituting the numerical values into eq. 6.8 we find:

$$\dot{m}_e = -L\left(4 \times 10^{-13}\frac{R}{M} + 10^{-11}\frac{\phi}{X}\right) \ M_\odot \ \text{yr}^{-1}. \tag{6.9}$$

Since ϕ/X is of the order of unity, we find that when $R/M \simeq 25$ (where R and M are given in solar units), the contributions of the two processes are equal. In an AGB with $m_c = 0.6M_\odot$, we have a radius of $R > 300R_\odot$. It is clear that mass loss from the stellar surface causes the main consumption of the envelope. With the increase in the radius and decrease in the mass, this ratio increases further. Thus we see that while the rate of increase in the core is proportional to the luminosity, the mass loss rate is greater than the increase rate of the core by more than an order of magnitude. In the AGB phase, the increase of the core is very small while the envelope is consumed by mass loss. Such a process causes stars in wide range of initial masses to converge to almost the same core mass when the envelope is entirely consumed. As mentioned in Section 6.4, the core mass will not be much larger than $0.6M_\odot$, irrespective of the initial stellar mass.[4]

In the case of stars with initial masses of 5 to 8 $M_\odot$, the envelope will be entirely consumed when the core is 1 to 1.3 $M_\odot$, which is still below the Chandrasekhar mass limit.

We have concentrated on the evolution of low-medium and medium mass stars. They make up the population for which the preceding description is relevant. Low mass stars cannot complete their lifetimes on the main sequence during the lifetime of the Galaxy. The evolutionary track of stars with high initial masses is entirely different and is treated separately.

6.8 Planetary Nebulae and the Creation of White Dwarfs

When the stellar hydrogen-rich envelope is entirely consumed, the bare core is left as a remnant of the star. If the core mass is around $0.6M_\odot$, the luminosity of the star toward the end of the AGB is about $7,000L_{\odot}$. When the envelope mass is a few hundredths of a solar mass, the radius of the star begins to shrink. Initially this process is slow, but with further decrease in the envelope's mass the rate increases. The rate of mass loss toward the end of the AGB is 10^{-6} to $10^{-5}M_{\odot}$ yr^{-1}, and the mass leaves the star as a regular stellar wind with a velocity of 20 km sec^{-1}. This velocity is on the order of magnitude of the escape velocity from the stellar surface of a one solar mass red giant.

When the radius shrinks, the mass loss rate increases for a short period by an order of magnitude. This short phase is called the *superwind phase*.[6] Throughout this stage the luminosity does not change; therefore when the radius decreases, the effective temperature increases, and the star moves to the left on a horizontal track in the H-R diagram. We now have to distinguish between the evolution of the two parts — the core and the remnants of the envelope. The evolution of the core is simple. During the course of a few thousand years the radius of the star shrinks to almost the radius of the core itself. The core then moves on a line of constant luminosity to the left of the H-R diagram and reaches an effective temperature of 120 to 150 thousand degrees. Later on it begins to cool, its luminosity decreases gradually, and it continues on the evolutionary track of white dwarfs. Figure 6.4 displays the evolution of the core as it changes to a white dwarf through an intermediate phase as a nucleus of a planetary nebula. Similar results were obtained by Schonberner.[6]

During this time, the remnants of the envelope continue to recede from the star as an expanding gas cloud. When the bare core reaches an effective temperature of 35,000 degrees, a very fast wind, probably created by the radiation pressure of the core, accelerates the matter that was last to leave the star. This matter catches up with the gas cloud and forms a shock front which proceeds outward and likewise forms the inner boundary of the gas cloud. The energies of the radiation photons become sufficiently high to ionize the hydrogen gas, which on recombination reradiates the ionizing energy in the wavelength of visible light. Most of the core radiation is in the range of ultra violet (UV) and is invisible to optical instruments. However the energy reradiated by the gas cloud appears as a spectacular shining nebula. An object in this stage is called a *planetary nebula* (PN), which is an expanding gas cloud shining by the reradiation of the core's energy. The core is now the central star (the nucleus, NPN) of the PN.

A detailed density profile of a PN named NGC 6826 (no. 6826 in the New General Catalog) was calculated by Plait and Soker.[7] The details are

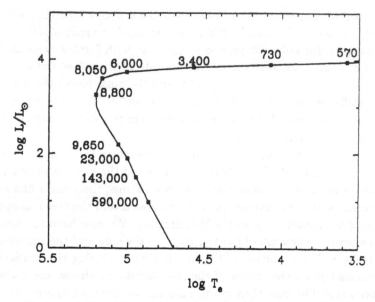

Figure 6.4. Evolution of the nucleus of a planetary nebula. The numbers along the line designate time (in years) elapsed from the end of the AGB.

displayed in fig. 6.5, in which the density of the nebula matter is given as a function of the distance from the central star (the radius of the nebula). The distance of this nebula from us is about 4,600 LY, and its height above the Galactic plane is about 1,000 LY. In the figure the radius of the nebula is given in arcsec, where at a distance of 4,600 LY each arcsec is equivalent to 0.02230 LY. For the density, the figure shows the electron density. Given that the hydrogen is fully ionized, the helium is singly ionized, and the nebula has a solarlike composition, a density of 1,000 electron per cm^3 is equivalent to a mass density of 2×10^{-21} gm cm^{-3}. The distance to the planetary nebula is uncertain to about 50 percent. This uncertainty is inherent in all the lengths cited in the paragraph below.

On inspection of fig. 6.5 we find that the inner boundary of the nebula is at a radius of 0.067 LY (3 arcsec) from the central star. From this boundary, the density increases to a maximum of 2.9×10^{-21} gm cm^{-3} at a radius of 0.115 LY (5 arcsec). Beyond this point the density drops, at first in a moderate slope and later in a steeper slope. The boundary between the moderate and the steeper slope is at a radius of 0.279 LY (12.5 arcsec). At a radius of 0.78 LY (35 arcsec) the density is very low, but still not zero,

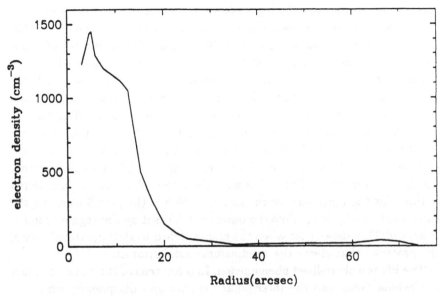

Figure 6.5. The density profile of NGC 6826. [Reproduced from Plait and Soker.[7]]

until it increases by a small value at a radius of 1.338 LY (60 arcsec) before vanishing at a radius of 1.561 LY (70 arcsec).

In order to calculate the mass enclosed in the different regions of the nebula, we integrate (numerically) the expression for a mass element:

$$dm = 4\pi r^2 \rho dr. \tag{6.10}$$

Integrating eq. 6.10 up to a radius of 0.279 LY (12.5 arcsec), we find that this part of the nebula has a mass of $0.15 M_\odot$. This point is the boundary between the moderate and the steep slope of the density. Actually, this is the boundary between the inner nebula and its halo. The density in the halo is indeed very low, but it extends over a huge volume which compensates for its low density. Continuing the integration up to the radius of 1.561 LY, we find another $0.39 M_\odot$ of mass, which brings the nebular mass up to $0.54 M_\odot$.

Noting that the outer regions of the nebula were the earlier ones to leave the star, we can relate the different regions in the nebula to the different types of stellar wind mentioned earlier. The fast wind which developed upon the unveiling of the hot bare stellar core formed the inner boundary of the nebula. This boundary, which is now located at a distance of 0.067 LY from the nucleus, was formed during a 400 to 1,000 year phase when this wind had a velocity of approximately 20 to 50 km sec^{-1}. The superwind probably formed the high density inner part of the nebula. With a velocity of about 15 km sec^{-1}, the wind reached the outer boundary of this region, at a distance of 0.279 LY, in about 5,000 years. The amount of mass included in this region, $0.15M_\odot$, demands a mass loss rate of about 3×10^{-5} $M_\odot$ yr^{-1}, which is a reasonable rate for this phase. The mass residing outside this point was ejected from the star by a regular wind, whose mass loss rate is lower by a factor of 10 than that of the superwind. The slight increase in density at a large distance from the nucleus (1.338 LY) is probably due to an interaction of the wind with the interstellar matter. It is worth noting that the inner part of the nebula, which is the part formed by the superwind, is elliptical, while the outer part formed by the regular wind is spherical. This shows that when the star switched to the superwind phase, the processes that create the wind assume axial symmetry.

The PN is a short-lived phenomenon. In a few tens of thousands of years the nebula fades into the interstellar medium and disappears, while the central star becomes a white dwarf, moving gradually along its cooling track in the H-R diagram. Soker[8] gives a comprehensive and illuminating description of a PN evolution.

References and Comments

1. Hayashi C., 1966, *Ann. Rev. Astron. Astrophys.*, **4**, 171.
2. Schonberg M., Chandrasekhar S., 1942, *Astrophys. J.*, **96**, 161.
3. We have several times repeated the assertion: "A temperature gradient is formed to transfer the energy produced. ..." It should be understood that this is not a planned program aimed at some ultimate goal. Instead, the creation of a temperature gradient to transfer an energy excess is a self-adjusting process. When an energy excess exists at a certain point, this point becomes hotter than its vicinity. The temperature difference between this point and its surroundings causes the energy to flow away from the location of the energy excess. When this excess is transferred away from this point, the point cools down and the energy flow stops. If an energy excess is produced continuously by a nonthermal source at a certain point, the temperature gradient between this point and its surroundings will self-adjust by the above process; the resulting temperature gradient will be exactly sufficient to transfer the excess energy at the rate it is produced. This is a steady-state configuration. The negative feedback existing between the temperature of a mass element and the decrease in its energy content by the energy flow is the "thermostat" of the system.
4. Weidemann V., Koester D., 1983, *Astron. Astrophys.*, **121**, 77.

5. Reimers D., 1975, in *Problems in Stellar Atmospheres and Envelopes*, ed.
B. Bascheck, W.H. Keggel, G. Tremain, Springer, Berlin.
6. Schonberner D., 1983, *Astrophys. J.*, **272**, 708.
7. Plait P., Soker N., 1990, *Astron. J.*, **99**, 1883.
8. Soker N., 1992, *Sc. American*, **266**, 5, 36.

Chapter 7

Binary Systems

Many stars exist in binary and even in multiple star systems. Only in systems that are close to us can we observe the separate components directly. Astrophysicists today believe that the number of binary systems is still greater than what we now observe. New binary systems are discovered to this day, even among bright stars whose existence we have known for long time. When the mass ratio between the companions is large, it is difficult to observe the binarity, although even very low mass companions can influence the system significantly. Many phenomena in stellar spectra that could not be satisfactorily explained in the past were later found to derive from binarity.

Most binary systems, such as spectroscopic and eclipsing binaries, are observed by indirect methods. Since observing the rotation period of a system and investigating stellar spectra do not depend on distance to the systems, we have been able to gather a great deal of information about such systems.

7.1 Types of Binary Systems

The important property of a binary system is the dynamic interaction between its components. The presence of the companion influences the dynamic behaviour of each component. There are a number of types of binary systems defined as follows:

Optical binaries. These stars happen to be in the same line of sight from Earth, but have no dynamic interconnection between them. They may even be at different distances from us.

Visual binaries. These stars are connected dynamically as a system. Such systems are close enough to us for observing both their components separately.

Astrometric binary systems. In such a system only one component is observed directly, but its behaviour shows that its motion is influenced by the presence of an unobserved component. By analyzing the motion of the observed component, we can investigate parameters of the system such as its period, its total mass, and the separation between its companions. From the spectral type and the luminosity of the observed component, we can deduce its mass; from this mass we can likewise deduce that of the unobserved companion as well. For example, in 1844 Bessel was able to deduce from the motion of the star Sirius (one of the brightest stars in the sky) that it is a component of a binary system with a rotation period of 49 years. Twenty years later the companion was observed, too, and found to be a white dwarf with a mass of about $0.9 M_\odot$. This system is today classified as a visual binary system.

Spectroscopic binaries. Although they are discerned to be a binary system from the Doppler shift in their radiation, these stars are either too close to each other or too far away from us for classification as visual binaries. We can calculate their velocities from the Doppler shift, and the velocities and the orbital frequency of the system provide us with much information.

Spectrum binaries. A Doppler shift cannot be observed for these stars, however the analysis of their spectrum reveals that there are two different sources with different characteristic spectra.

Eclipsing binaries. These stars cover each other for a certain period of time during their orbital rotation, and a significant dip in the luminosity can be observed due to the eclipse.

Plainly the observation of binary systems depends to a large extent on the angle between their plane of rotation and our line of sight. The Doppler shift in the spectrum owing to the orbital velocities and the eclipsing of the companion are best observed in systems whose rotation planes are aligned edgewise in our direction. Visual binary systems are best observed when the system's rotation plane is aligned face-on in our direction. Probably the alignment of the planes of rotation for these systems is randomly distributed in all directions, and we must take into account that our observations yield only the projection of the real configuration on a plane perpendicular to our line of sight.

7.2 The Physics of a Binary System

The basic physics of the system derives from the gravitational interaction between the components. Two objects that attract each other gravitationally will stay in a periodic rotational motion if the centrifugal force created by their rotation balances the gravitational attraction. The two components rotate with the same frequency around their common centre of mass. The distance of each component from the centre of mass is inversely proportional to the component's mass. In a coordinate system whose origin is located at the system's centre of mass we have:

$$m_1 r_1 = m_2 r_2 \qquad (7.1)$$

where subindices 1 and 2 denote the two components, and r_1 and r_2 are distances of the components from their centre of mass, given in absolute values. This equation yields for the distances ratio:

$$\frac{r_1}{r_2} = \frac{m_2}{m_1}. \qquad (7.2)$$

In circular motion the orbital velocity v_1 is given by $v_1 = r_1 \omega$, where ω is the orbital frequency. Since this frequency is the same for the two components, the velocities ratio is the same as the distances ratio.

When we consider a circular rotation, the centrifugal force acting on m_1 always equals the gravitational force acting on the same mass:

$$m_1 \omega^2 r_1 = \frac{G m_1 m_2}{(r_1 + r_2)^2}, \qquad (7.3)$$

and a similar equation holds for m_2:

$$m_2 \omega^2 r_2 = \frac{G m_1 m_2}{(r_1 + r_2)^2}. \qquad (7.4)$$

We divide eq. 7.3 by m_1, and eq. 7.4 by m_2, add the two equations, and multiply by $D^2 = (r_1 + r_2)^2$ to obtain:

$$\omega^2 D^3 = G(m_1 + m_2) = GM. \qquad (7.5)$$

Here M denotes the total mass of the system, and D denotes the separation between the components. This equation yields Kepler's third law for the planets. In the solar system, M is dominated by the solar mass and D is practically the distance of the planet from the centre of mass of the system, located very near to the solar centre. Hence Kepler found that for the planets there is a constant ratio between the square of the period,

P $(P = 2\pi/\omega)$ and the third power of the distance of the planet from the Sun. The expression derived from eq. 7.5 for the orbital period of a binary system is:

$$P = \left(\frac{4\pi^2 D^3}{GM}\right)^{1/2}. \tag{7.6}$$

This period is the clearest observation time for a binary system. If the system is near enough for us to observe the separation directly, we can calculate the total mass of the system. If the system is aligned edge-on in our direction, we can observe the velocities of the components by the Doppler shift in their radiation. The frequency is observed independently of distance. By using the relation $v_1 = r_1\omega$, we find the distances of the components from the centre of mass along with their separation. Using eq. 7.5, we can calculate the total mass from the separation and the frequency. Since the mass ratio of the components is the inverse of the velocities ratio, each of the masses can be found as well. If the system is not aligned edge-on in our direction, what we actually observe is the projection of the velocities on a plane aligned in our direction. Thus our observations and calculations yield only lower limits for the variables.

The dynamic calculations of a binary system are actually the only direct way by which we can determine the mass of stars. The first few hundred binary stars observed furnished much information about the distribution of stellar masses. Necessarily, these stars were also close enough to Earth for their spectrum and the other parameters to be seen clearly. From this information, the researchers could deduce the statistical properties of the distribution, and find relations such as the mass-luminosity relation presented in fig. 1.1.

7.3 An Example

The use of eqs. 7.2 to 7.5 to determine the parameters of a binary system can be illustrated by following Herbig and Moorhead's investigation[1] of the binary system designated as BD-21^{0}6267A. This system is itself the brighter component of the visual binary system BDS 11854, which has a separation of 24 arcseconds. The system BD-21^{0}6267A is found to be a spectroscopic binary with a period of 4.083 days.

Figure 7.1 shows the spectrum of this system taken at certain intervals along the binary period. The clearly observed lines are the H and K lines of twice ionized Ca, and several lines of hydrogen. The different spectra shown in the photograph are denoted by the fraction of the phase at which they were taken.

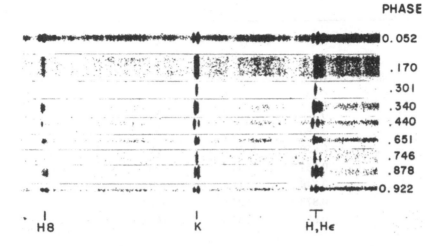

Figure 7.1. Photograph of the spectrum of BD-21°6267A. [Adopted from Herbig and Moorhead.[1]]

In most of the spectra we clearly see that each individual line is composed of two lines, one of them stronger and the other weaker. Moving from the upper spectrum downward, we move in time along the period of the system. The relative locations of the two lines interchange twice during a period; in the transition between the opposite locations, the two lines coincide. During a complete period, the two lines interchange twice and recover their initial location. The location of a line denotes its wavelength. Thus we see that each line undergoes an elongation and a shortening of its wavelength during each period, and this process occurs in opposite directions in the two lines.

The change in wavelength is due to the Doppler shift caused by the relative motion of the sources along our line of sight. The wavelength grows shorter when the source moves toward us, and becomes longer when the source moves away from us. We assume that we have two sources (stars) which move in circular orbits. By calculating the exact shift of the wavelength, Herbig and Moorhead determined the velocities of each component as they change along the period. The stronger lines are related to Star 1, and the weaker lines pertain to Star 2.

Table 7.1 displays the velocities as a function of the fraction of the period. The velocities are given in km sec^{-1}. Positive velocities denote velocity toward the observer.

Table 7.1.

Phase	v_1 (Star1)	v_2 (Star2)
0.052	+34.5	−65.8
0.109	+29.0	−53.2
0.301	−21.3	. . .
0.340	−36.3	+20.6
0.460	−53.5	+49.1
0.560	−52.5	+42.3
0.651	−36.4	+25.7
0.746	−11.8	. . .
0.878	+23.6	−52.6
0.937	+33.5	−60.6

We observe that the distribution of the velocities is not symmetric with respect to null velocity. This means that the system as a whole is in motion relative to us. Taking averages of the velocities, we find that the system moves away from us at a velocity of about 9 km sec^{-1}. This is probably due to its motion in the wider visual binary system BDS 11854. Correcting for the velocities to obtain those of the stars in their system's centre of mass, we find that the maximum velocity of Star 1 is 46.8 km sec^{-1}, and that of Star 2 is 58.1 km sec^{-1}. Recall that the observed velocities are the projections of the real velocities on the plane aligned edgewise in our direction. We do not know the angle between this plane and the plane of rotation of the system. We have to consider $r_1 \sin i$ instead of r_1, where i is the angle between our line of sight and the axis of rotation.

Using the relation between the distance of the star from the centre of mass and the star's observed velocity ($r \sin i = v/\omega$), we find $r_1 \sin i = 2.63 \times 10^{11}$ cm $= 3.76 R_\odot$, and $r_2 \sin i = 3.26 \times 10^{11}$ cm $= 4.66 R_\odot$. They yield the separation $D \sin i = 8.42 R_\odot$. The angular velocity, ω, found from the period, is $\omega = 1.78 \times 10^{-5}$ sec^{-1}. Using eq. 7.5, the total mass of the system, M, can be found from ω and D. The ratio of velocities yields the masses ratio, from which each separate mass can be found. Since D in eq. 7.5 appears with the third power, the masses are found, multiplied by $\sin^3 i$, and we have: $m_1 \sin^3 i = 0.271 M_\odot$, and $m_2 \sin^3 i = 0.218 M_\odot$.

This example demonstrates how the measure of the orbital frequency and the Doppler shift in the component's radiation yield a complete picture of the system. It is easily seen that the value of i, the angle of inclination of the rotation axis to our line of sight, has considerable significance in all these observations. This value cannot be measured directly. However, we can make an estimate by considering whether the system is eclipsing or not. If it is, we have to consider the nature and degree of the eclipse. An eclipse which presents a flat bottom in the minimum light shows that the eclipse is total. Otherwise, if the system does not eclipse, a statistical weight for this value is accounted for in considering observational results.

Kepler's law (eq. 7.3) provides a way to use the period of a binary system in characterizing the separation and mutual interaction between its components. A system with a separation indicating a period longer than few years is one in which the only mutual influence is the dynamic behaviour reflected in the component's orbital motion. No significant influence has been found which makes the evolution of the components different from the evolution of single stars.

7.4 Tidal Effects

For a system in which the orbital period is less than one year, the separation between individual stars is shorter than the distance from the Earth to the Sun. During the main sequence evolution, the components act by tidal forces on each other, which for closer binaries might have a significant influence on the behaviour of the stellar surface.

Tidal deformation occurs in each component of the system owing to the gravity of the other component. The tidal deformation remains at the same location on the star if the axis of revolution of the star is perpendicular to the plane of rotation of the system, and if the frequency of the stellar revolution is the same as the orbital frequency. This tidal disturbance revolves with the star and is always directed toward the other component. The stellar spin is synchronous with the orbital revolution. In such a situation there is a minimal loss of energy resulting from tidal distortion. (For instance the Moon remains in such a position relative to the Earth, and the same half of the Moon is always facing earthward.) When binary systems form, this situation is not usually the case, and the stars spin at different frequencies (ordinarily higher) than the orbital frequency of the system. As a result of the viscosity and friction in the star, the tide tries to remain in the direction of the other component and is partially dragged with the stellar revolution, like the tides on Earth. The friction causes a loss of energy and angular momentum which is manifested in a deceleration of the stellar spin. The stellar spin now becomes increasingly synchronous with the orbital revolution.

Observation has found that in many close binary systems the orbits of the companions are not circular and reveal a significant eccentricity. Such systems are certainly too young to have completed the synchronization. The equation which connects the eccentricity, ε, to the orbital energy and the angular momentum of the system is:

$$\varepsilon^2 = 1 + \frac{2M}{G^2 m_1^3 m_2^3} E J^2 \tag{7.7}$$

where J is the angular momentum, and E is the energy of the system. (Recall that $\varepsilon^2 = 1 - \frac{b^2}{a^2}$, where a and b are the semimajor and the semiminor axes of the ellipse, respectively.)

When energy is lost from the system, the right term on the right-hand side of eq. 7.7 decreases, and when it equals -1, the eccentricity vanishes. When the stellar spin synchronizes with the orbital frequency, they obtain a steady-state. The conditions for steady-state are:

(1) That the orbit is circular ($\varepsilon = 0$).

(2) That the axes of revolution of the stars are perpendicular to the plane of rotation of the system.

(3) That the spin of each of the components is synchronous with the orbital frequency. The time scale for synchronization is usually much shorter than for the circularization of the orbit. By calculation, we find that in a steady-state the orbital angular momentum should be greater than the sum of the spin angular momenta of the two components. If this condition is not met, the system will not remain in a steady-state. The stars will then either spiral inward to merge or spiral outward to a greater separation. The initial energy of the system determines which of these two alternatives occurs.

7.5 The Common Envelope

When a star in such a system becomes a red giant, its radius may increase to some hundreds of solar radii; and as it expands, the star may absorb its companion into its envelope. Usually the two components are not of the same mass. The more massive star evolves faster and becomes a red giant, while the other remains a main sequence star. If the red giant loses mass, as in a continuous stellar wind, there is a chance that part of the lost mass will be accreted by the companion. If a significant amount of mass is accreted in this way, it enhances the companion's mass, thus altering the companion's subsequent evolution. Ultimately, if the second star survives this stage, it will later become a red giant, and the role of the partners in the interaction may interchange.

When the envelope of a red giant star engulfs its main sequence companion, the evolutionary track of the companion may take an interesting route. The red giant's envelope is very dilute. The companion continues its orbital motion inside this envelope, so it has now actually become a common envelope for both stars. The companion loses kinetic energy and angular momentum as a result of friction with the envelope matter. Its orbit becomes smaller, and it spirals inside the envelope toward the core of the red giant.[2] During this motion, it may also accrete mass from the envelope through which it is moving. In other cases, the companion may lose mass and gradually evaporate into the envelope.

The fate of such a system depends on the conditions in which these processes take place and on the exact configuration of the red giant: whether it is an AGB star, how much mass it contains in its envelope, and at what rate it loses mass by a stellar wind. The two processes — the loss of mass from the envelope by a stellar wind and the inward spiralling of the companion — compete with each other. The companion may actually reach the core of the red giant before the envelope is consumed. In such a case the companion will be accreted to the core and enhance its mass. If the companion contains hydrogen-rich matter, this event will be explosive because the hydrogen will start to burn violently when the companion hits the burning shells at the core's edge.

On the other hand, the envelope mass may be entirely consumed before the companion reaches the core. In such a case the red giant turns into a white dwarf star. The companion is no longer exposed to friction and may continue orbiting the other star in a much closer orbit. The overall result is a close binary system. This scenario is one of the possible ways in which close binary systems are formed.

7.6 Close Binary Systems

When the two components in a binary system are very close to each other, individual mass elements in both companions "feel" the gravitation of the other star. When we plot the equipotential surfaces of the gravitational potential for the two close objects, we find that near the centre of the system, the equipotential surfaces of each are similar to those of a single object. When we move further away from the centre, however, the equipotential surfaces of each object are disturbed by the gravitational potential of the other, until we arrive at an equipotential surface that touches the equipotential surface of the other object. Equipotential surfaces beyond this surface are no longer separate for each object, but shared by both.

If we move far away from the system, the common equipotential surfaces will approach spheres, like equipotential surfaces of a single source composed of the two objects. When the equipotential surfaces of the two objects touch each other, the gravitational attraction of the two objects is equal. If a mass element from one object moves and passes this point to the other side, it will be attracted to the other object.

The preceding description is correct if we consider only gravitational potential. However, in a close binary system, the rotational velocity is sufficiently high so as to balance the gravitational attraction; and the centrifugal force must be considered as well. In a co-moving system of binaries (which is a coordinate system that rotates with the binary system), the centrifugal effects may be represented by the centrifugal potential, Φ_c, at a given point, x, as:

$$\Phi_c = -\frac{1}{2}\omega^2 d^2 \tag{7.8}$$

where ω is the angular velocity of the system, and d is the distance from the axis of rotation to the point x. In order to formulate the problem, let us use a coordinate system whose origin is at the centre of mass of the system, and D is the separation between the two masses m_1 and m_2. Axis x_1 passes through the centres of the two objects, and axis x_2 is perpendicular to x_1 in the plane of rotation. Axis x_3 is perpendicular to the plane of rotation. The expression for the total potential Φ, at a point x, is composed of the gravitational potentials of the two objects and the centrifugal potential:

$$\Phi = \Phi_{G1} + \Phi_{G2} + \Phi_c = -\frac{Gm_1}{q_1} - \frac{Gm_2}{q_2} - \frac{1}{2}\omega^2 d^2 \tag{7.9}$$

where q_1, q_2 are the distances from x to m_1 and m_2 respectively. Consider the case in which x is located in the plane of rotation. Then $q_1 = [(x_1 - r_1)^2 + x_2^2]^{1/2}$; $q_2 = [(x_1 - r_2)^2 + x_2^2]^{1/2}$, and $d^2 = x_1^2 + x_2^2$. (Recall that in the coordinate system we are using, either r_1 or r_2 is negative.) In addition, we use the relation between ω and D, (eq. 7.5): $\omega^2 = GM/D^3$. Substituting this values for q_1, q_2, D and ω in eq. 7.9, we have:

$$\Phi = -\frac{Gm_1}{[(x_1 - r_1)^2 + x_2^2]^{1/2}} - \frac{Gm_2}{[(x_1 - r_2)^2 + x_2^2]^{1/2}} - \frac{GM}{2D^3}(x_1^2 + x_2^2). \tag{7.10}$$

On the equipotential surfaces, Φ is constant. The force in the system, f, is found by the gradient of Φ:

$$f = -\nabla\Phi. \tag{7.11}$$

It is clear that the negative gradients of the gravitational potential terms in Φ are negative, and they represent the gravitational attraction of the objects. The negative gradient of the centrifugal potential is positive. It represents an outward directed force called the centrifugal force. The addition of the centrifugal potential changed the system in such a way that with increasing distance from the centre of the system, the centrifugal force overcomes the gravitational attraction. Close to the system centre, therefore, the sum of the forces yields an attractive force, and matter is bound. But matter which moves further out from the system centre will become unbound and will be repelled from the system. The intersection of the equipotential surfaces with the plane of rotation form equipotential lines. In fig. 7.2 we display the equipotential lines in the plane of rotation of a system in which m_1 is greater by a factor of three and a half than m_2.

We treat eq. 7.10 as an equation for x_1 and x_2 for a given value of Φ. We are especially interested in the solutions of eq. 7.10 for which $\nabla\Phi = 0$.

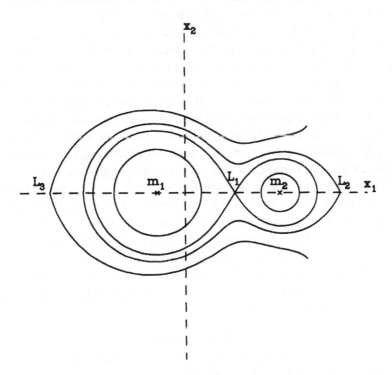

Figure 7.2. Equipotential surfaces in the plane of rotation in a binary system in which $m_1 = 3.5 \times m_2$.

The points that fulfill this condition are those at which matter does not feel any force in any direction. They are points of extremum (or actually a local maximum) in the specific potential energy of the system.

Equation 7.10 is a complicated one, and we shall not solve it here in detail. For $\nabla\Phi = 0$, it has five roots in the plane of rotation, meaning that there are five points at which no force acts on the matter. These points are called *Lagrangian points*. Three of these points, designated in the figure by L_1, L_2 and L_3 respectively, are located on the line connecting the centres of the two objects. L_1 is located between the two objects. L_2 and L_3 are on either side of m_1 and m_2. L_4 and L_5 are located on a line perpendicular to x_1, and each forms an equilateral triangle with the centres of m_1 and m_2. A mass element located at L_1 is in an equilibrium between the attractive forces of the two objects, and any deviation in either direction will cause it to accelerate in that direction. A mass element located on any of the other four Lagrangian points is in equilibrium between attraction and repulsion. A small deviation inward from these points will cause it to accelerate inward, while a small deviation outward will cause it to accelerate outward.

The volume enclosed in the equipotential surface containing the point L_1 is called the *Roche-lobe*. In the present case we have such a lobe for each of the objects. The matter inside this surface is bound to the object located in that lobe. If the matter overflows the boundary of the Roche-lobe, it will leave the object and either accrete to the other object or eject from the system, depending on the point at which it crossed the Roche-lobe boundary. Later we shall discuss in detail several examples in which mass transfer between the two components causes significant phenomena.

An interesting instance is when one of the companions possesses high radiation pressure, which actually acts to decrease the effective gravitational force. Suppose that in the case described in fig. 8.2, the radiation pressure of Star 1 is significant. This means that the effective gravitation of this star is smaller, and L_1 moves away from Star 2 and toward Star 1. If Star 2 is filling its Roche-lobe, it will not shed mass through L_1, which has receded, but rather through L_2. If this mass is still bound to the system, it will start a spiral motion around the binary system and probably approach Star 1 in a complicated orbit.

References

1. Herbig G.H., Moorhead J.M., 1965, *Astrophys. J.*, **141**, 649.
2. Soker N., 1991, *Astrophys. J.*, **367**, 593.

Chapter 8

Star Formation

Star formation takes place in large molecular clouds with sufficiently high density and sufficiently low temperature to allow the gravitational collapse of protostellar objects. There is a great difference between the conditions needed for the formation of low mass stars (with masses up to $2M_\odot$) and those required for the formation of higher mass stars. Let us begin by considering the formation of low mass stars.

A typical giant molecular cloud[1] has a mass of about $10^5 M_\odot$, a radius of about 60 LY, and a mean molecular density of 50 molecules cm^{-3}. Such clouds seem to contain lower order clouds of 10^3 to $10^4 M_\odot$, with a radius of 5 to 10 LY and a density of few hundred molecules cm^{-3}. The temperature in such a cloud is about 10 degrees K.

We have shown in eq. 6.1 that if the cloud density is uniform, the inward acceleration due to gravity is proportional to the distance from the centre of gravity. This means that in a radial free-fall of matter particles, the contraction is homologous — that is, the ratio between the distances of different parts of the cloud and the centre is constant. This homologous contraction is disturbed when the particles' velocities are converted to heat and the pressure starts to play a role in the balance of the forces. But the process of the cloud contraction to form a star is not straightforward. We shall try to describe some of its stages in detail. A striking observation is the "inefficiency" of star formation. From a molecular cloud with a mass of $10^4 M_\odot$, several hundred stars form with masses of about one solar mass each, while the rest of the cloud disperses in the interstellar medium. This process yields an efficiency of a only few percent in the use of raw material.

Before going on to describe these stages, let us first consider two phenomena connected with the process of star formation: T Tauri stars (T Tau) and Herbig-Haro objects (HH).

8.1 T Tauri Stars

T Tau stars are those similar to the one designated by "T" in the constellation Taurus. Designating a star in a group by a capital letter means that it is a variable star. The variability of T Tau stars is irregular. The characteristics of these stars are as follows:

(1) They are always found close to bright or dark nebulae, which are clouds of gas and dust.

(2) They vary irregularly in both their luminosity and spectral properties.

(3) Their light includes bright emission lines of hydrogen and calcium.

(4) Their effective temperature is low, at around 4,000 degrees K.

(5) They have a high abundance of lithium, indicating that they are young stars.

(6) Their surface shows a very active behaviour: matter winds, flares, spots, etc.

The typical radius of a T Tau star is three times the solar radius. The spectrum of part of these stars has an infrared excess relative to the characteristic spectrum of stars with the same effective temperature. We believe that this infrared excess is a result of radiation from the dust which surrounds the star, and emission from the accretion disk which resides around the stellar equator.

The high lithium abundance in T Tau stars and the intrinsic connection between these stars and clouds of dust suggests that we are witnessing the first appearance of newly born stars. A molecular cloud with a mass of $6 \times 10^3 M_\odot$ covers the area around the constellations Taurus and Auriga, and contains about 100 T Tauri stars. Since this cloud is relatively close to the solar system (at a distance of about 450 LY), it is the object of intense observation with the aim of studying and characterizing these stars.

8.2 Herbig-Haro Objects

The astronomers Herbig and Haro independently observed these objects in the early 1950s. They are blobs of matter moving at high velocities through the interstellar medium and are usually found in pairs with the two components moving away from each other in opposite directions. HH1 and HH2 were the first discovered, and immediately scientists began looking for the "exciting star," under the assumption that two objects receding from each other in a straight line probably emerged from a common source. It was only 30 years later that they observed the exciting star. The reason for the delay was that the source is a very cool object having an effective temperature of only a few hundred degrees; its radiation therefore is in very long wavelengths which cannot penetrate the Earth's atmosphere. When telescopes were finally mounted on satellites, observations from above the Earth's atmosphere revealed vital information concerning these objects. IRAS (Infra-Red Astronomical Satellite), which was launched in January 1983 obtained the main body of this information during a period of 300 days. The IRAS observations used the wavelengths 12, 25, 60, and 100 μm, and covered about 96 percent of the sky. These observations illuminated the particulars of these cold objects, and among them were several exciting stars of HH objects. HH objects are likely jets of matter that emerge in the opposite direction from very cool objects believed to be protostars. The velocities of HH objects relative to the interstellar medium are 200 to 300 km sec^{-1}. Their distances from each other are a few light years. About 100 HH objects have been observed to date.

A fine system of HH objects was observed[2] at a distance of about 430 LY from Earth. The exciting star of this system is Th28, from which two HH objects stream on both sides in opposite directions. Observers believe that this system exhibits the highest degree of symmetry yet known for any comparable system. The two HH objects, named HHE and HHW (HH East and West), are at distances of 0.06 LY and 0.072 LY from the source respectively. Their velocities are about 300 km sec^{-1}. The observed luminosity of the exciting star is only 0.015 $L_\odot$, and it is likely that a highly opaque disklike cloud obscures the star. The nice symmetry of the system suggests that the medium in which the matter jets are moving is highly homogenous. Figure 8.1 displays the two objects, HHE and HHW, on both sides of Th28. Two matter jets with lengths of 0.024 and 0.027 LY emanate from Th28 in the direction of the HH objects. The scale in the figure is in arcseconds (1″), where at the distance of the system each arcsecond corresponds to 2×10^{-3} LY.

The observed feature of an HH object is actually the shock front formed at the interface between the streaming jet and the interstellar medium. In some HH objects, several such features follow one another in the same direction. The objects HH7, HH8, HH9, HH10, and HH11 form such a train of consecutive shock fronts, moving in the same direction with velocities of 170 km sec^{-1} away from their source, SSV13. It appears that there were several consecutive bursts of material jets from the exciting star.

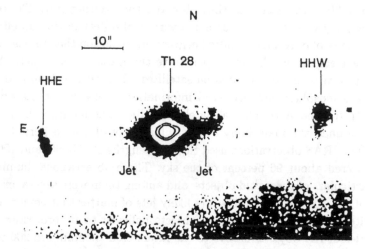

Figure 8.1. Th28 with the two HH objects, HHE and HHW. [Adopted from Krautter.[2]]

The analysis of the radiation from the cool exciting stars shows that they are rotating, and that the jets emerge from their poles. The counterparts of some of the HH objects cannot be observed, probably because they are obscured by the dusty cloud through which they are streaming.

8.3   Stages in Star Formation

The formation of low mass stars in a molecular cloud occurs in the following four stages:[3]

(1) The formation of cloud cores.

(2) The gravitational collapse of the cloud core to form a protostar and an accretion disk.

(3) The bipolar flow stage, which creates the HH objects with their exciting stars.

(4) The revelation of the star as a T Tauri star with remnants of a nebular disk.

The Cloud Core

A cloud of matter is subject to the mutual gravitational attraction between its constituents driving the system toward collapse. If there were no

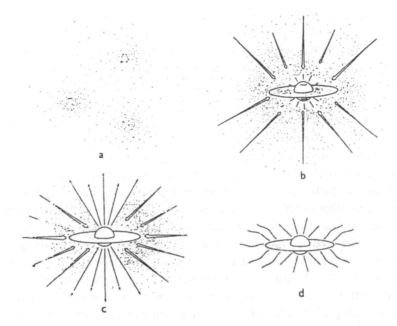

Figure 8.2. The four stages of star formation. (a) Cores form within molecular clouds as magnetic and turbulent support is lost through ambipolar diffusion. (b) A protostar with a surrounding nebular disk forms at the centre of a cloud core collapsing from inside-out. (c) A stellar wind breaks out along the rotational axis of the system, creating a bipolar flow. (d) The infall terminates, revealing a newly formed star with circumstellar disk. [Adopted from Shu, Lizano, and Adams.[4]]

mechanism to support the cloud against such a collapse, we would expect all the clouds to collapse on a dynamic time scale and star formation to conclude within a very short time. However, when the cloud contracts to form a protostar, the contraction is a slow process which advances in an almost quasistatic manner. During this phase, a cloud core forms surrounded by an envelope. The core is a loose structure with a uniform temperature and a density which falls outward as $1/r^2$. The envelope surrounding it is composed of dilute gas and dust. The inefficiency of star formation is also connected with this low rate of star formation.

The quasistatic character of the cloud collapse indicates that some mechanism exists which impedes the process and prevents a free-fall collapse from taking place. Three mechanisms were "suspected" as possible candidates for this role: turbulent movements within the cloud matter; rotation of the cloud or of portions of it; and the presence of magnetic fields.

Observations show a uniform distribution of the magnetic fields over the entire cloud. This situation limits the scale of possible turbulent motions, so that no global turbulence exists in the cloud. Calculations of turbulent velocities and their possible influence show that the effect of turbulence could not furnish support against the collapse of the cloud.

The idea that rotation might play this role assumes that when a sample of matter is contracting, the conservation of angular momentum dictates that the contracted object will rotate with higher angular velocity than its angular velocity in the uncontracted state. This increase compensates for the decrease of the rotation radius. The process delays contraction until the particles of matter can lose most of their angular momentum so they can be accreted to the central zone. The angular velocity of a typical cloud is about 3×10^{-14} rads sec^{-1}. Observations have shown that cloud cores rotate with the same low angular velocity as their surrounding envelopes. The centrifugal forces created by such an angular velocity are too weak to take on the function of supporting the cloud against rapid collapse.

The possibility that therefore remains is the action of magnetic field in the matter. The strength of the magnetic field in the interstellar medium is a few microgauss (μG). When the matter begins contracting, the magnetic flux is conserved. Because of the higher density of the contracted matter, the density of the magnetic flux increases and the magnetic field grows stronger. In a molecular cloud, a magnetic field of few tens μG is possible. Further contraction leads to a further increase in the magnetic field. However, the mutual interaction between nearby portions of parallel magnetic fields acts like a pressure opposing the development of a denser magnetic flux or a stronger magnetic field. Thus the contraction of magnetized matter creates "magnetic pressure" which opposes the contraction. The cloud thereby reaches a state of equilibrium, in which the "magnetic pressure" and gravitational attraction balance each other out. Further contraction can take place only when the magnetic field is weakened by some mechanism.

What can such a mechanism be? A magnetic field cannot influence neutral particles. Only charged particles in motion are subject to magnetic interaction. Most of the material in the molecular cloud is neutral. A certain fraction of the particles is ionized by cosmic rays traveling through the Galaxy, and by natural radioactivity existing in the matter. These ions and the free electrons are subject to the influence of the magnetic field. The neutral particles move toward the centre of gravity in response to the gravitational attraction. During their motion they encounter friction with the charged particles, which are "held" against such a contraction by the magnetic field. This friction couples the bulk of matter to the magnetic field, and the contraction of the cloud is slowed to a gradual drift of the matter's neutral component, instead of proceeding as a dynamic collapse. The magnetic field thus delays the contraction, causing it to become a quasistatic process. Owing to this drift, the neutral particles concentrate at the centre without carrying magnetic flux with them. The magnetic flux at the core becomes less dense, and the global effect is a leakage of the magnetic field outward.

Quantitatively, the relation between the gravitation and the delay effect of the magnetic field is represented by the expression for the critical mass, m_{crit}, which is the limit for a flux to mass ratio that can still support the cloud against collapse:

$$m_{crit} = 0.15 \frac{\Phi}{G^{1/2}} = 10^3 \left(\frac{B}{30\mu G} \right) \left(\frac{R}{6LY} \right)^2 M_\odot. \qquad (8.1)$$

Here Φ is the magnetic flux, G is the gravitational constant, B is the magnetic field, and R is the radius of the object. B is given in units of $30\mu G$, which is believed to be the typical field in such clouds. If the mass of the object is supercritical $(m > m_{crit})$, the object will collapse. If the mass is subcritical $(m < m_{crit})$, the contraction of the object will be a slow quasistatic process. Clouds with supercritical masses are the sites of formation for high mass stars. In subcritical mass clouds, only low mass stars are formed.

The magnetic field is also responsible for the core and its envelope rotating at the same angular velocity. The field lines of the magnetic field are continuous over the boundary between the core and the envelope. A differential rotation along the boundary gives rise to a shear in the field lines which opposes the differential rotation. Thus the magnetic field acts as a "magnetic brake" on the core, causing it to corotate with the envelope.

The action of the magnetic field both in delaying contraction and in braking the core leads to a loss of energy and weakening of the magnetic field. Thus the general picture is of an increase in the core from the drift of the neutral particles toward the centre, and of a decrease of the magnetic flux as a result of the leakage of the magnetic field.

In the overall process, we obtain an increased core and a weakened magnetic field, until a point of instability is reached, at which the quasistatic contraction ends and a collapse of the core begins. In clouds with temperature of about 10 degrees K, the quasistatic phase lasts for about 10^6 yr.

The second stage begins with the collapse of the core.

Gravitational Collapse of the Cloud Core

At the beginning of this stage, the energy released by the contraction radiates in the form of infrared radiation. Since the opacity of the neutral matter is low, the radiation reaches the surface immediately and the core is almost isothermal. When the accreted matter becomes opaque to this radiation, a temperature gradient develops. At a radius of 3×10^{10} cm, a shock front forms where the free-falling matter encounters the denser contracted core. This core becomes a protostar wherein gradients of density, temperature, and pressure form. The inward motion of matter continues, and, because of the rotation, a rotating disk forms around the equator of the protostar. The mass contained in the disk is about 10^{-6} to $10^{-5}M_\odot$. The coupling between the core and its envelope by the magnetic field lines is eliminated when the core collapses. Differential rotation then develops in which the core rotates with higher angular velocity than the disk.

As a result of collisions in the disk, the infalling matter loses kinetic energy and angular momentum and is accreted onto the protostar, whose mass increases. This matter accretes with high entropy as it heats up in consequence of collisions in the disk or passage through the shock front. Meanwhile the central temperature increases to above a million degrees, which is the threshold for deuterium burning. The small amount of deuterium contained in the matter burns and supplies energy in addition to that released by the contraction. The calculations show that for an accretion rate of $10^{-5}M_\odot$ yr^{-1}, deuterium burning begins when the protostar's mass is about $0.3M_\odot$.

At this stage the opacity inside the protostar is high. A convective zone develops at the centre and gradually advances outward. The radiation which reaches the protostellar surface from the interior encounters the infalling matter and imparts radiation momentum to the matter particles. In the beginning, the infalling matter overcomes the matter repelled outward, and the accretion continues over the whole stellar surface. With the increase of the convective zone, the luminosity reaching the surface from the stellar interior increases. With it, the amount of outward momentum imparted to the matter particles increases as well. This process continues until the matter streaming outward plows its way through the cover of infalling matter at its weak points. The disk, which forms a thick impenetrable cover, blankets the equator and the low altitudes around it. The weak points are the two poles, where the cover is thinner and the matter moving outward breaks through in the form of bipolar jet streams. The third stage, at which the Herbig-Haro objects are formed, begins with this bipolar flow. The convective zone reaches the surface when the stellar mass is about $0.5M_\odot$.

The Bipolar Flow

In the third stage, the protostar is the exciting star of the HH objects. The transition to this stage depends heavily on the deuterium burning. Convective transport is needed for the transfer of energy created by burning of the deuterium. Near the stellar surface, the interaction between the convection flow and the differential rotation results in a dynamo for energy drift toward the stellar surface. The energy creates a matter wind that flows outward from the surface of the star. The wind begins at the poles as a bipolar flow, gradually widening to increasingly cover more of the stellar surface. The process of revealing the stellar surface does not depend on the way in which the matter is accreted to the star — either through an accretion disk or directly over the entire surface. It depends instead on physical processes taking place in the stellar interior: the energy production by nuclear reactions and convective transport of that energy.

The amount of mass that can be accreted to the star depends on the rate of the accretion. Higher rates allow an accumulation of bigger masses before the deuterium ignition, which is followed by the formation of stellar wind. With accretion at a rate of $10^{-7}M_\odot$ yr^{-1}, the stellar wind begins when the star has a mass of $0.1M_\odot$. A rate of $10^{-6}M_\odot$ yr^{-1} yields a wind when the stellar mass is $0.3M_\odot$; a rate of $10^{-5}M_\odot$ yr^{-1} yields a wind when the stellar mass is $0.5M_\odot$; and a rate of $10^{-4}M_\odot$ yr^{-1} yields a wind when the stellar mass is $2M_\odot$.

When the bipolar flows begin, the accretion of matter continues and the inward and outward flow of matter occur simultaneously. With the increase in stellar wind, the bipolar flows become wider and disrupt the cover on greater portions of the star. The stellar wind actually cleanses the stellar surroundings of the cover by repelling matter and ejecting it back to the interstellar medium. Here we see the main cause of the inefficiency of star formation. Part of the matter already added to the star is ejected by the stellar wind. Yet much more matter returns to the interstellar medium, even before it reaches the star.

The revelation of the star is possible after the stellar wind has "washed away" the cover of infalling matter. It is at this moment that the protostar appears as a star. During all the preceding stages, the star moved down a vertical line at the right side of the H-R diagram (Hayashi track). It becomes observable only when the stellar wind cleanses its surroundings. This is the moment of "star birth." Stahler[5] has calculated the location

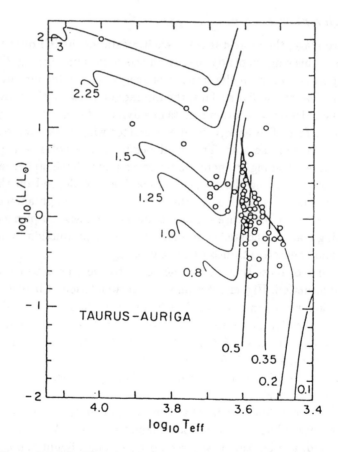

Figure 8.3. Birthline (marked by a heavy curve in the figure) and the location of a group of T Tau stars, in the H-R diagram. The lines marked by numbers display the evolutionary track toward the main sequence of stars with the corresponding masses. [Adopted from Stahler.[5]]

of this point along the evolutionary tracks of stars of various mass in the H-R diagram. These points form a line in the diagram which intersects the Hayashi tracks for different mass stars. This line is called the "birthline" of stars. When stars pass this line during their descent along the Hayashi track, they appear as T Tau stars. Figure 8.3 shows the location in the H-R diagram of a group of T Tau stars from the Taurus-Auriga cloud. Nearly all of the stars are close to and below the birthline sketched in the figure.

T Tau Stage

At the T Tau stage, the stellar wind flows from the entire stellar surface. Remnants of the disk surround the equator. The star rotates slowly because of the magnetic braking that took place during the preceding stages. The influence of the disk on the observed radiation of the star depends on the disk activity. A passive disk contains cold matter that may re-radiate energy originally emitted by the star and absorbed by the disk. An active disk contains hot matter heated by the collision of the disk matter with the accreted matter. Such a disk radiates energy released in these collisions. For the creation of an active disk, the rate of mass accretion must be high. With low rates of mass accretion, the disk will be passive. An active disk contains 0.1 to 1 $M_\odot$, which may later accumulate and form a secondary star. A passive disk contains much less mass, and this matter may possibly form planets.

8.4 Formation of High Mass Stars

The process described above is responsible for the formation of low mass stars with a stellar mass of up to $2M_\odot$. The formation of higher mass stars demands different conditions. We spoke earlier of the existence of supercritical and subcritical mass clouds. In subcritical mass clouds, only low mass stars can be formed. High mass stars are formed in supercritical mass clouds. Such clouds are usually much bigger and are characterized by the presence of ionized hydrogen matter (called HII zones) in their vicinity.

How are supercritical clouds formed? The lifetime of supercritical clouds is short because they collapse, and star formation in such a cloud takes place on a short time scale. The fact that such clouds exist today shows that they form continuously. A typical example is the big cloud observed in the sword of Orion, where there are many young high mass stars. This cloud is at a distance of 1,500 LY from us. Its dimensions are 100 LY, and its mass is about 10^5 $M_\odot$.

When two subcritical clouds merge, this process can take place in different ways. For example, if the magnetic fields of the clouds are parallel and collide in a direction perpendicular to the field lines, then the unified cloud formed by this collision will have a higher mass and a higher magnetic flux. The ratio of mass to flux will remain the same as that of the colliding clouds, and the new cloud will be subcritical as well. If, on the other hand, the collision takes place along the mutually shared field lines, the mass of the new cloud will double, but the flux will remain the same. The ratio of mass to flux will increase. Several consecutive collisions will form big clouds whose mass to flux ratio will become supercritical.

Collisions of the second type occur most frequently in zones of ordered motions of clouds, such as those in the Galaxy's spiral arms or in a collison of two galaxies. The spiral arms of the Galaxy are not constant structures. They are formed continuously by density waves traveling in the Galaxy, eventually fading into the Galactic disk as a result of the tidal forces. The density waves form in the Galactic disk by the differential rotation of the Galaxy. They rotate in the Galaxy as spiral shock fronts. Zones of high pressure and density, created behind the advancing shock fronts, form the spiral arms. The density excess of the spiral arms above the average density of the disk is not high. However, since these zones become the formation sites of brilliant luminous stars, their presence stands out against the background of the Galactic disk. The lifetime of the spiral arms is short, and the formation of high mass stars takes place during this period.

The collapse of the supercritical cloud is swift. However it is still delayed by magnetic pressure relative to free-fall collapse. A protostar begins to form at the centre of such a collapsed region. When the amount of accumulated matter is close to one solar mass, the centre becomes sufficiently hot for nuclear reactions of hydrogen burning to begin. The object becomes a star, while matter continues to flow in from the surroundings and is accreted to it. The star becomes bigger and heavier. The more it gains in mass, the hotter its centre and the higher its rate of life become. What we have now is a mature star that continues to grow in mass resulting from the infall of matter from its surroundings. The collision of infalling matter with the stellar surface heats the outermost layer of mass. The outer zone expands, the stellar radius increases, and luminosity now radiates by a low effective temperature. When such a star reaches a mass of 8 to 10 $M_\odot$, its luminosity is a few thousand times the solar luminosity. Its effective temperature is then a few tens of thousands of degrees. The radiation at this effective temperature is mainly in the optical and the UV range of wave-lengths. Such high luminosity, radiated at so high an effective temperature, has a radiation pressure which exerts a force exceeding the gravitational attraction on the infalling matter (recall eq. 2.12). When this matter is exposed to the radiation pressure, it will be repelled into the interstellar medium.

At the beginning of this stage a cocoon of cold matter still surrounds the star. This matter absorbs the short wavelength radiation and re-radiates it in the form of infrared radiation whose photon energies are much lower than that of the UV. Although radiation pressure at the stellar surface exceeds the Eddington limit, the inflow of matter and mass accretion to the star continues until the cocoon of cold matter is disrupted by the hot radiation. The infalling matter is exposed to the radiation pressure of the UV radiation. At this moment the mass accretion ends, and the matter in the stellar surroundings accelerates away from the star. This is when the star becomes a young luminous star. Stars of this type are called O stars and have outflow winds with velocities of a few thousand km sec^{-1}. The newly born star now disrupts and destroys the surrounding cloud from which it was formed.

The upper limit on a stellar mass of 60 to 80 $M_\odot$ probably derives from the fact that at such high masses, the star matures and develops significant radiation pressure before attaining higher mass. The radiation pressure terminates the mass accretion. Thus, the limit on the stellar mass derives from processes caused by the high mass of the star.

Some features of the formation track are similar in the two modes of star formation: the formation of low mass and of high mass stars. These features consist of the partially delayed collapse, the mass accretion that continues after the stellar core forms, the revelation of the star resulting from the "cleansing" of its surroundings by stellar wind, etc. The main difference between the two modes is the rate at which each stage in the formation takes place. These rates determine how much mass accumulates at each stage, and the scale of the final product, the star. The formation of high mass stars takes place on shorter time scale than of low mass stars. The rapid collapse of the supercritical cloud yields higher mass cores, which become very hot, and reach the threshold for hydrogen fusion. Instead of deuterium burning in low mass stars, which supplies only moderate amounts of energy, there is a full chain of hydrogen burning which furnishes high luminosity. A high effective temperature is needed for the radiation of this luminosity. In this way radiation pressure creates the stellar wind. Due to the collapse of the supercritical cloud, which continues until the radiation pressure of the star disrupts the cloud, the rate of mass accretion is higher by a few orders of magnitude than that of cores in subcritical clouds. The amount of accreted mass is accordingly high.

The significant energy production in the star begins only after the accretion of more than one solar mass. The continual high accretion rate implies that large amounts of mass accumulate until the star is revealed. The consequence of this chain of events is that in large supercritical clouds having a high rate of mass accretion, the final product will be a high mass star of 8 to 40 $M_\odot$. However, the formation of high mass stars involves a self-regulating mechanism in the form of radiation pressure that terminates the mass accretion before very high masses accumulate.

8.5 Formation of the Solar System

The general track for the formation of the Sun is the formation of a low mass star. However the possibility of observing many details and studying relics from the formation epoch in the present solar system put strong constraints on the theories of stellar evolution.

The present distribution of angular momentum in the solar system is such that 99 percent of the system's total angular momentum is stored in the orbital motions of the planets (the two giant planets, Jupiter and Saturn possess 80 percent of this angular momentum). The Sun itself stores 99.86 percent of the mass of the system. This distribution is due to the way in which the system was formed from the progenitor cloud.

When a molecular cloud contracts to form a star, the angular momentum content of the cloud demands a high rotational velocity of the contracting object, unless some mechanism causes a loss of angular momentum. We have already seen how the magnetic braking mechanism acts to transfer angular momentum from the central contracting zone to the outer expanding zones of the star. Creating a binary system, in which a large amount of angular momentum is stored in the orbital motions of the companions, absorbs angular momentum. The majority of the stars in the vicinity of the Sun exist in binary systems.

A direct dynamic collapse usually creates elliptical orbits with nonvanishing eccentricity. Here the progenitor cloud breaks into two (or more) fragments which later form the companion stars. Clouds containing high values of angular momentum tend to form binaries by direct dynamic collapse. Another alternative for creating a binary system is one in which a companion (usually the companion having the lower mass) forms from the disk which surrounds the central core of the cloud. Systems formed by such a scenario have fairly circular orbits. In such a case the mass range of the secondary (or secondaries) can be wide. An extreme case may be one in which the secondary's mass (or the secondaries' masses) is on the order of magnitude of a planet. It is likely that the solar system formed by a disk scenario.

A number of recent observations support the validity of the disk scenario:

(1) Several main sequence stars were found to possess a thin disk around their equator. Usually, such a disk is characterized by the excess of infrared radiation in the stellar luminosity. Most of these observations were made by IRAS. One example is the star Vega in the Constellation Lyra.[6] Vega is a young MS star (its estimated age is 3×10^8 yr), with a mass of $2M_\odot$ an effective temperature of 9,700 degrees, and luminosity of $56L_\odot$. Its distance from the solar system is about 26 LY.

The analysis of the data shows that the radius of the thin disk around Vega is about 85 times the distance of the Earth from the Sun, and it is composed of particles whose dimensions are of the order of few millimetres (compared to grains in the interstellar medium whose dimensions are of the order of 10^{-3} to 10^{-4} mm). The temperature of the disk matter is 85 K, and the mass contained in the disk is estimated to be few hundred times the Earth's mass($M_\oplus$). Vega is much younger than the Sun and the dimensions of the grains in its disk may represent a certain stage in growth of the grains in the process of planet formation.

These observations may be the first evidence for the existence of a planet formation process outside the solar system.

(2) The star HD114762, which resides at a distance of 90 LY from us, has a very low mass companion, where the companion's mass is about 10 times the mass of Jupiter.[7] Such low mass objects (around $0.01M_\odot$) are too cool to ignite nuclear reactions and are called "brown dwarfs." Their luminosity is too low to be observed directly. The observations are made by measuring the changes in the radial velocities of the massive companion in the system, and such a system is considered as an astrometric binary.

It is found that the changes in the radial velocities of the star HD114762 have a period of 64 days, and the mass of the unseen companion is about $0.01M_\odot$. Theoretical calculations show that such a low mass object cannot be formed by a fragmentation of the collapsing cloud when the star is formed, and hence the disk scenario is preferred.

(3) PSR 1257+12 is a 6.2 millisecond pulsar residing at a distance of 1,600 LY from us.[8] It possesses two planets, whose masses are $2.8M_\oplus$ and

$3.4M_\oplus$. Their distances from the pulsar are are $100R_\odot$ and $77R_\odot$, and they orbit the pulsar in almost circular orbits with periods of 98.2 and 66.6 days respectively. Millisecond pulsars are recycled pulsars that were spun up by a high rate mass accretion from their former companion (see Section 12.4). It is suggested that part of the mass that approached the pulsar formed a disk, which later turned into planets.

Obviously, planets that orbit a pulsar cannot have the substantial conditions needed for the development of any kind of life, but the existence of these planets shows that the formation of planets from a thin accretion disk is not a rare possibility.

The nature and composition of meteorites produced in the early epoch of the solar system give an indication of the conditions that existed during this period. An interesting feature in the composition of meteorites is the presence of products from the nuclear decay of ^{26}Al. The half-life of this radioactive element is 7.4×10^5 yr, meaning that the meteorites formed not later than few times 10^5 yr after the original material accreted to the cloud. It was suggested that this element was produced in a supernova and injected somehow into the protosolar cloud. However, such a scenario would have yielded additional results that were not observed. Radioactive elements like ^{26}Al are produced in the burning shells of AGB stars. These elements might be dredged up to the stellar surface and ejected into the interstellar medium by a stellar wind or through the ejection of a planetary nebula. This is probably the way in which this material reached the protostellar cloud.

A type of meteorite which is called *chondrite* includes *chondrules*. Chondrules are circular objects with a diameter of about one millimetre that reside in a grainy matrix. These objects were probably produced from the fusion of the material in the matrix under special conditions of pressure and temperature. A theory for the formation of the solar system should explain how chondrules were formed.

Mineralogically, it is clear that the chondrules formed at a high temperature and cooled within a few minutes of formation. Some chondrules exist inside other chondrules, meaning that the inner chondrule formed earlier, and the outer one later. This situation shows that the production of chondrules was a continuous process and not a one time event. Remnants of magnetism found inside the chondrules show that during the epoch of their formation the magentic field was of 0.1 to 10 Gauss.

Another important feature of meteorites is that many of their constituents were formed in the presence of a high ratio of oxygen to hydrogen.

This ratio is higher by two orders of magnitude than that found in the solar composition. Likely, sedimentation of oxygen-rich objects, such as solid grains or snowflake-like aggregates, throughout the outer part of the solar nebula caused this oxygen enrichment. Thus the inner part of the nebula became oxygen enriched, while the outer part retained its original high hydrogen content. This may have implications for the differences in composition between the inner and the outer planets.

The Primitive Solar Nebula[9]

When the primordial cloud collapses, it contains a nucleus surrounded by a nebula, the primitive solar nebula. As a result of rotation the nebula flattens to a disk. Due to the viscosity in the disk matter, there is friction between adjacent rotating layers, which spin with different velocities. This friction heats the matter, converting kinetic energy to thermal energy. The disk matter becomes luminous, and the excess heat flows outward either by radiation or convection.

There are three stages in the evolution of the nebula. Stage I is the stage of the disk formation, during which infalling matter from the interstellar medium accumulates on the disk. During this stage the density and temperature of the disk increase. Stage II is a steady-state stage, in which the rate of mass accretion to the disk equals the rate of mass transfer from the disk to the central nucleus. This is the main dissipation stage. During Stage III the mass accretion from outside gradually vanishes, and the mass of the disk decreases as a result of accretion of mass to the nucleus. The nucleus grows to form the central star, the Sun.

Stage I lasts for about 10^4 yr. At its conclusion, the radius of the star is approximately $15R_\odot$. The beginning of the star formation is the moment at which hydrostatic equilibrium between gravity and the pressure gradient halts the collapse at the centre. The volume reaching this equilibrium increases gradually until the end of Stage II. The Sun passes the birthline in the H-R diagram with a radius of about $4R_\odot$, which occurs at the end of Stage II.

The Formation of the Planets

The gas which makes up the nebula usually includes small aggregates of matter. Such an aggregate is composed of a core formed while the matter was still in a stellar atmosphere and an envelope accreted to the core. When the core formed, the gas from which it was made was in a state of thermal equilibrium. The envelope, on the other hand, accreted when the gas was cooler and dilute. This envelope is a mixture of atoms and molecules which encountered the core in the interstellar medium. When the matter in the nebula becomes denser, these aggregates interact with each other. The relative velocities between the aggregates are low, and they stick to each other when they collide. They may form larger aggregates up to the scale of one centimetre, even before they accrete to the disk. Such aggregates may be the raw material for the production of chondrules.

The disk is not a homogenous medium. It contains gas, grains, and larger aggregates, each responding differently to the conditions in the disk. Three types of forces act on each particle in the disk: the gravitational force, the centrifugal force, and thermal pressure. The acceleration imparted to any object by gravitation and by the centrifugal force is exactly the same for all objects. The acceleration imparted by the pressure gradient, however, depends on the ratio of the object's surface area to its mass. This ratio decreases with increasing mass of the object. Thus if the gas particles are in equilibrium between the three forces, larger objects such as grains and aggregates are influenced less by the pressure gradient pushing outward, and they drift toward the centre. The larger the object is, the greater its drift. As a result of this drift, which is orthogonal to the rotational motion of the gas, the aggregates accrete more matter, until they attain dimensions of metres and kilometres.

Gradually the disk matter decouples into two zones: an inner zone, where the large aggregates prevail, and an outer zone, where the gas prevails. The estimated time needed for the creation of aggregates of dimensions of one kilometre is about 10^4 yr.

Further increase in the aggregates by collisions caused by the differential drift velocities leads to the formation of objects with dimensions of a few kilometres. These are the constituents of the asteroid population. Calculations of the statistical rate of formation for larger objects show that this ratio obeys approximately the law $dN/dm \simeq m^{-0.8}$, where N is the number of objects with mass m.

Very large objects may go into a "runaway" accumulation. This can happen when the velocity of a small encountering object is lower than the escape velocity from the larger object. This statement may be formulated by the effective cross-section of the large object, σ_{eff}:

$$\sigma_{eff} = \sigma_{geo} \left(1 + \frac{v_e^2}{v^2} \right) \tag{8.2}$$

where σ_{geo} is the geometrical cross-section of the large object, v is the velocity at infinity of the encountering object, and v_e is the escape velocity. The ratio v_e/v may become very large thus causing a high rate of accumulation. This is probably the way in which the inner lower mass planets formed, with masses on the order of about 10^{27} gm and radii of a few thousand kilometres.

A special question concerns the formation of the giant planets, Jupiter and Saturn. (The masses of Jupiter and Saturn are 318 and 95 times the mass of the Earth, respectively.) The explanation of their formation is that once a massive core with a mass of 10 to 20 $M_\oplus$ forms, it will induce gravitational instabilities in the ambient gas envelope and drive it to collapse on the massive core. Indeed model configurations of these planets show that they possess a core composed of heavy elements with a mass of 10 to 20 $M_\oplus$. The question still remains of how these cores formed. The giant planets were formed in the outer part of the solar nebula, where the oxygen to hydrogen ratio is not enhanced by sedimentation and the abundance of oxygen is lower in these planets than in the inner planets. The composition of the inner planets includes a high fraction of heavy elements, that form the iron core of the Earth, and the stoney crust on which we live. The relics of hydrogen and helium left in the atmosphere of the inner planets, were ejected to the interstellar medium, because their thermal velocities are higher than the escape velocity from these planets. The composition of the big, outer planets includes mainly light elements, and these planets do not possess solid crust.

References

1. Lizano S., Shu F.H., 1987, in *Physical Processes in Interstellar Matter*, ed. G.E. Morfill, M. Scholer, Reidel Publishing Company, Dordrecht.
2. Krautter J., 1987, in *Circumstellar Matter*, ed. I. Appenzeller, C. Jordan, Reidel Publishing Company, Dordrecht.
3. Shu F.H., Adams F.C., 1987, in *Circumstellar Matter*, ed. I. Appenzeller, C. Jordan, Reidel Publishing Company, Dordrecht.
4. Shu F.H., Lizano S., Adams F.C., 1987, *Ann. Rev. Astron. Astrophys.*, **25**, 23.
5. Stahler S., 1983, *Astrophys. J.*, **274**, 822.
6. Aumann H.H., Gillet F.C., Beichman C.A., De Jong T., Houck J.R., Low F.J., Neugebauer G., Walker R.G., Wesselius P.R., 1984, *Astrophys. J. Lett.*, **278**, L23.
7. Mazeh T., Latham D.W., Stefanik R.P., Torres G., Wasserman E., 1990, in *Active Close Binaries*, ed. C. Ibanoglu, Kluwer Academic Press, Dordrecht.
8. Wolszczan A., Frail D.A., 1992, *Nature*, **355**, 145.
9. Cameron A.G.W., 1988, *Ann. Rev. Astron. Astrophys.*, **26**, 441.

Chapter 9

Rotation of Stars

Most of the analysis of stellar evolution used nonrotating models. This choice was made for the sake of convenience. Although we know the equations, exact solutions are very difficult to acheive. However we do know that all astrophysical objects rotate, and the rotation undoubtedly affects the evolutionary tracks of these objects to some extent. This means that the consequences deduced from the analysis of nonrotating objects only suit objects with a low rotation velocity. Here we consider the effects of rotation as small perturbations on the zero order approximation, the nonrotating model. The earliest introduction of rotation effects into stellar evolution calculations was actually through calculating rotation-induced perturbations on nonrotating calculated models.

The effects of rotation are most important when big changes take place in the stellar radius. In such a case, the conservation of angular momentum results in severe restrictions on the angular velocity of the stellar surface. When the star undergoes rapid contraction, as when its configuration changes from that of a red giant to that of a horizontal branch star (after helium ignition in the core), the surface of the contracted star should rotate at a much higher angular velocity than the extended star. During the contraction, the star usually eliminates the excess angular momentum through mass loss with high specific angular momentum. This might have implications for the character of the stellar wind at this phase.

On the other hand when a red giant forms, its envelope expands and its core contracts. As a result of the conservation of angular momentum, the stellar surface will rotate much more slowly, and the core will rotate much faster. As a result a differential rotation develops. A high degree of differential rotation results in important effects such as a stretching of magnetic field

151

lines, creation of turbulent motions, energy losses, and redistribution of the angular momentum through radial and meridional mass flows in the star.

When observing the spectral lines of a rotating star, the lines originating in the receding part of the star are red-shifted. The spectral lines originating in the approaching part of the star are blue-shifted. All stars, apart from the Sun, are observed as point sources, and the different velocities at opposite sides of the rotating star cannot be measured separately by their Doppler shift. However both effects are observed, yielding a broadening of the spectral lines of the star. The extent of broadening depends on the magnitude of the rotational velocity and the inclination of the axis of rotation to our line of sight. The rotational velocity at the equator, v_e, causes maximum broadening. The parameter inferred from the broadening of the spectral lines is $v_e \sin i$, which is the projection of v_e on our line of sight. The inclination angle between the axis of rotation and the line of sight is i.

In this chapter we describe several methods for treating rotation in stars. We shall also discuss some of the astrophysical cases in which this treatment may be of importance.

9.1　The Coordinate System

When treating a nonrotating model we have spherical symmetry. The only coordinate which parametrizes the stellar structure is r, the distance from the centre of the configuration. Such models are called *one-dimensional models*. In rotating objects, the models deviate from spherical symmetry, and we have a symmetry only around the axis of rotation. To parametrize such a system we must use two coordinates, where the additional coordinate parametrizes the changes taking place along the axis of rotation. Usually a pair of coordinates is used: either r and θ, where r is the distance from the centre and θ is the angle measured from one of the poles; or z and ξ, where z is the coordinate along the axis of rotation and ξ is the orthogonal distance from this axis.

For a point with given r and θ, the z and ξ coordinates are given by:

$$z = r \cos \theta ; \quad \xi = r \sin \theta.$$

With the r, θ coordinate system we deal with patches whose depth is dr, and whose area is $r^2 d\theta \sin \theta d\phi$. Due to axial symmetry, the expression does not depend on the azimutal angle ϕ, and we deal with circular patches with area of $2\pi r^2 \sin \theta d\theta$, and depth dr. With the z, ξ coordinate system, we deal with hollow cylinders with height dz, a perimeter of $2\pi\xi$, and a depth of $d\xi$.

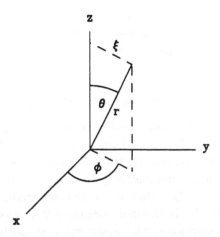

Figure 9.1. The coordinate systems used to present rotating stars.

When the angular velocity, Ω, is a function of ξ only, $\Omega = \Omega(\xi)$, $d\Omega/dz = 0$, we can infer the effects of rotation from a *centrifugal potential*. We use an *effective potential*, which includes the centrifugal potential, Φ_c, in addition to the usual gravitational potential, Φ_G:

$$\Phi = \Phi_G + \Phi_c \qquad (9.1)$$

where Φ is the effective potential. The gradient of Φ gives the *effective gravitational acceleration*, g_{eff}, which includes the centrifugal acceleration in addition to the proper gravitational acceleration:

$$g_{eff} = -\nabla\Phi. \qquad (9.2)$$

The centrifugal potential Φ_c, at a given distance ξ from the axis of rotation, is:

$$\Phi_c = -\frac{1}{2}\Omega^2\xi^2 = -\frac{1}{2}\Omega^2 r^2 \sin^2\theta. \qquad (9.3)$$

The gravitational potential, $\Phi_G(x)$, at a given point x, is given by:

$$\Phi_G = -G \oint \frac{\rho(x')}{|x - x'|} dV(x'). \qquad (9.4)$$

The integration is carried over x'. In the z, ξ system $|x - x'|$ is given by: $(x - x')^2 = (z - z')^2 + \xi^2 + \xi'^2 - 2\xi\xi' \cos\phi$, where ϕ is the azimutal angle difference between the radius vectors to ξ, and to ξ'. We see clearly that for a spheri-symmetric matter distribution, eq. 9.4 yields the familiar expression: $\Phi_G = -\frac{Gm(r)}{r}$.

We are looking for a method which will take care of the deviations from spherical symmetry. We shall describe two such methods: the *self-consistent field* method, and the method of the *Eulerian coordinate system*.

9.2 The Self-Consistent Field

In this method we treat barotropic models or barotropes. A barotrope is a configuration in which the isopicnic surfaces (surfaces with constant density) coincide with the isobaric surfaces (surfaces with constant pressure). Polytropes are such configurations, where the pressure is a function of the density only, $P = P(\rho)$. Due to the thermodynamic relations, the same surfaces will also be isothermal. Temperature, density, and composition determine the evolution of the stellar composition by nuclear transmutations. Hence starting with uniform composition, the surfaces of constant composition will also coincide with the isopicnic surfaces, and these surfaces will therefore be equipotential surfaces. The gradient of the pressure is orthogonal to the equipotential surface, as is the effective gravitational force, which includes the gravitational force plus the centrifugal force. The energy flux by radiation and convection is proportional to the temperature gradient, which is also orthogonal to the equipotential surfaces. We thus find that the equipotential surfaces play the same role here as in the spherical symmetric configuration, except that in this case they are not exact spheres. If we keep the masses between two neighbouring equipotential surfaces constant, we have constant mass shells with the directions of both energy flow and the pressure gradient orthogonal to the boundaries between the mass shells.

The calculation of the model structure can proceed almost in the same way as in the spherical symmetric model, when we use the mass shells as the units for the calculation, and the mass as the Lagrangian coordinate of the model. This is the *self-consistent field method.*[1] Models in which a more general form for the pressure exists in the form of $P = P(\rho, T, \dots)$, but in which the condition $d\Omega/dz = 0$ still holds, are called *pseudobarotropes*. We find that in a pseudobarotrope the equipotential surfaces exist as well, so that we can also use the method of the self-consistent field. As it turns out, even the quasistatic method which is described in Chapter 5 for spherical symmetric models can be used here. We calculate the energy flow between the mass shells according to the temperature gradient and the opacity at the boundary between the shells.

The condition $d\Omega/dz = 0$ implies that the specific angular momentum j

does not depend on z, and we write $j = j(\xi)$:

$$j(\xi) = \Omega(\xi)\xi^2. \tag{9.5}$$

In an object rotating with differential rotation but for which the barotropic (or pseudobarotropic) condition still holds, the centrifugal potential Φ_c at a given ξ is expressed by:

$$\Phi_c = -\int_0^\xi \frac{j^2(\xi')}{\xi'^3}\, d\xi'. \tag{9.6}$$

The model is characterized by a given distribution of the angular momentum as a function of ξ. We begin the calculations by guessing a density distribution for the model. From this distribution plus the given distribution of the angular momentum, the effective potential is calculated by using eqs. 9.4 and 9.6.

The right-hand side of the hydrostatic equation (eq. 2.2) represents the force (per unit volume) acting in the system. Since we represent the force here as the gradient of the potential, we write the hydrostatic equation as:

$$\nabla P = -\rho \nabla \Phi \tag{9.7}$$

or in a more specific form:

$$\frac{1}{\rho} dP = -d\Phi. \tag{9.8}$$

In a barotrope we have $P = P(\rho)$, which when substituted in eq. 9.8 yields an equation for ρ as a function of Φ:

$$\rho = \rho(\Phi). \tag{9.9}$$

(For a pseudobarotrope, we keep the entropy in each mass shell constant through this treatment, as we did in Section 4.3 of Chapter 4, and we can still use eq. 9.9.)

Using eqs. 9.8 and 9.9, we calculate a new density distribution from the potential calculated by eqs. 9.4 and 9.6. A new potential is calculated from this new density. These steps are repeated. By alternately calculating the density and the potential inferred from this density, we approach the correct model until obtaining a satisfactory measure of precision. The location of the equipotential surfaces as function of the mass, inferred from the density distribution, represents the configuration of the model. Models of rotating stars can be calculated by this method even for very high distortions caused by the effects of rotation.

The main difficulty of this method is that because of the lack of the spherical symmetry, the calculation of the proper gravitational potential is much more complicated than in the spherical case. We perform the calculations using an approximation technique.

The two conditions that must be maintained are the constancy of the mass shells and the condition $d\Omega/dz = 0$. These conditions imply that mass flow in the model cannot be handled by this method, and the flow of angular momentum is limited as well. We shall see later when these features are important.

9.3 The Eulerian Coordinate Net

If we do not want to be limited by the conditions imposed by the method of the self-consistent field, we use a coordinate system which does not depend on the model. Such a system is the *Eulerian coordinate net*, which is a coordinate system external to the model. (For sake of convenience, we can allow the coordinate system to expand or contract with the stellar radius, keeping the internal division of the coordinate mesh unchanged.)

We can use either pair of coordinates: r, θ, or z, ξ. In the case of each choice, the pair of coordinates defines a mesh of cells independent of the model. A flow of mass, energy, and angular momentum can take place between the cells, according to the equations of motion solved in each shell.[2]

The calculations in this method are much more complicated than in the previous one. For each cell, we calculate the flow of energy and mass in four directions, across the four boundaries of the cell along the two relevant coordinates. The gradients of the density, the composition, and the temperature across the boundaries between the cells determine the magnitudes of the flows.

Practically speaking, this is a dynamic method of calculation similar in its principles to the one described in Section 5.6 of Chapter 5. Consider an element of mass which has an initial entropy s_0, and an initial specific angular momentum j_0. We follow its motion according to the equations of motion. By following all the mass elements which compose the object, we obtain the evolution of the object in time. This method is very exact, and demands a huge amount of CPU time and computer memory for the calculations.

9.4 Meridional Circulation

Let us investigate whether a radiative equilibrium can be maintained in a rotating star.

The effective potential, Φ, which includes both the gravitational and the centrifugal potentials, is given in eq. 9.1. Here Φ_c is expressed either by eq. 9.3 or by eq. 9.6. The gravitational potential satisfies the relation:

$$\nabla^2 \Phi_G = 4\pi G \rho. \tag{9.10}$$

In a barotrope or pseudobarotrope, Φ_c does not depend on z, and from eq. 9.6 we obtain:

$$\nabla^2 \Phi_c = -\frac{1}{\xi}\frac{d}{d\xi}(\Omega^2 \xi^2). \tag{9.11}$$

Thus we have:

$$\nabla^2 \Phi = 4\pi G \rho - \frac{1}{\xi}\frac{d}{d\xi}(\Omega^2 \xi^2). \tag{9.12}$$

Now we calculate the radiation flux from an equipotential surface. Energy flows in the direction of the temperature gradient, which is orthogonal to the equipotential surface and parallel to g_{eff}. The radiation flux, H, is given by:

$$H = \frac{-4ac}{3}\frac{T^3}{\kappa\rho}\nabla T = \frac{-4ac}{3}\frac{T^3}{\kappa\rho}\frac{dT}{d\Phi}\nabla\Phi = f(\Phi)\nabla\Phi. \tag{9.13}$$

On the right-hand side of eq. 9.13, $f(\Phi) = \frac{-4ac}{3}\frac{T^3}{\kappa\rho}\frac{dT}{d\Phi}$ involves variables which depend on Φ alone, and the expression is thus constant for equipotential surfaces. Since the stellar surface is also an equipotential surface, the radiative flux from the stellar surface is proportional to $\nabla\Phi$. The gradient is constant for an equipotential surface only in spherically symmetric configurations. It cannot be constant for one in a star which is distorted from spherical symmetry, because the distances between adjacent equipotential surfaces become smaller as we move from the equator to the poles. On a given surface, g_{eff} at the poles is larger than it is at the equator. Thus in a distorted star the stellar luminosity varies over the stellar surface. We will consider the observational implications of these variations when we deal with Be-stars.

In an equilibrium state, H changes as a result of energy production by nuclear reactions. Thus we may write for a change in H, ∇H:

$$\nabla H = \frac{df(\Phi)}{d\Phi}(\nabla\Phi)^2 + f(\Phi)\nabla^2\Phi = \rho g \tag{9.14}$$

which is the condition for radiative equilibrium. g is the rate of energy production per unit mass by nuclear reactions. ∇H should vanish in the envelope of the star where no nuclear reactions take place. Substituting for $\nabla^2\Phi$ from eq. 9.12, we obtain:

$$\frac{df(\Phi)}{d\Phi}(\nabla\Phi)^2 + f(\Phi)\left[4\pi G\rho - \frac{1}{\xi}\frac{d}{d\xi}(\Omega^2\xi^2)\right] = \rho g. \qquad (9.15)$$

In the simple case of a uniformly rotating star, Ω is constant and

$$\frac{1}{\xi}\frac{d}{d\xi}(\Omega^2\xi^2) = 2\Omega^2.$$

Equation 9.15 reduces to:

$$\nabla H = \frac{df(\Phi)}{d\Phi}(\nabla\Phi)^2 + 4\pi G\rho f(\Phi)\left[1 - \frac{\Omega^2}{2\pi G\rho}\right] = \rho g. \qquad (9.16)$$

All the terms in eq. 9.16 are constant for an equipotential surface except the term $\nabla\Phi$. Therefore the coefficient of $(\nabla\Phi)^2$ in eq. 9.16 must vanish: $\frac{df(\Phi)}{d\Phi} = 0 \Rightarrow f(\Phi) = $ constant.

This means that $g \propto 1 - \frac{\Omega^2}{2\pi G\rho}$. An actual star obviously cannot fulfill such a condition. In the case where Ω is not constant but depends on ξ, we cannot so easily observe the conclusion. Nonetheless in a pseudobarotrope, Ω is constant on cylinders of constant ξ, while $\nabla\Phi$ is constant over surfaces which are not far from spherical. In order to satisfy eq. 9.15, we should demand that $\frac{df(\Phi)}{d\Phi} = 0$, and $\frac{1}{\xi}\frac{d}{d\xi}(\Omega^2\xi^2) = $ constant. Again we find a condition which cannot be fulfilled in an actual star.

The conclusion is that radiative equilibrium cannot be strictly maintained in a star distorted by rotation. This is called the *Von Zeipel paradox*. What are the implications of such a paradox?

The inference is that the temperature and pressure cannot be constant over the equipotential surfaces in a distorted configuration. The energy flux emerging from the stellar surface, which is proportional to g_{eff}, is lower at the equator and higher at the poles. Thus the effective temperature should also be lower at the equator and higher at the poles. A pressure gradient forms between these two areas causing a flow of matter along the meridians called *meridional circulation*. This flow is complemented by circulation in a plane passing through the axis of rotation. This circulation carries energy which balances the variations caused by ∇H.

In a steady-state of a spherically symmetric configuration, eq. 9.15 holds locally at each point. In a distorted star, the equation holds for equipotential surfaces in its average value only. The amount of energy transferred between adjacent mass shells confined by equipotential surfaces equals the average value derived from eq. 9.15 for the surface. Variations from this average exist locally at each point on the equipotential surface. Certain regions on such a surface transfer more energy than the average energy transfer calculated for the surface while others transfer less. In the outer parts of the star, where the temperature is too low for nuclear reactions, the average value of ∇H vanishes. A local increase in H causes local variations in heat content for which the heat carried by the meridional flow compensates. Thus the velocity of the meridional flow should satisfy the equation:

$$\nabla H = -\rho T \left(\frac{dS}{dr} v_r + \frac{1}{r} \frac{dS}{d\theta} v_\theta \right). \qquad (9.17)$$

In this equation, S is the entropy per unit mass, TdS is the energy content of a unit mass, and v_r and v_θ are the radial and meridional components of the flow, respectively.

Inspection of eq. 9.16 shows that for a uniformly rotating star, when the nuclear reactions vanish (and the first term on the left-hand side of the equation equals zero), the point at which $\Omega^2 = 2\pi G\rho$ is a critical point. At this point the sign of ∇H is inverted. This inversion of the sign leads to a reversal in the direction of the meridional flow. There are two regions of meridional flow which circulate in directions opposite to each other. Such a situation is displayed in fig. 9.2.

The relevant parameter in such calculations is the ratio of the centrifugal to the gravitational acceleration at the equator, $\alpha = \frac{\Omega^2 R^3}{GM}$. Usually we expect α to be a fraction on the order of thousandths to tenths.

The magnitude of the velocity for the circulation is $v_c \simeq \alpha \frac{LR^2}{GM^2}$. A characteristic time for this motion, t_c, is the time needed for a mass element moving with the velocity v_c to cross a distance equal to the stellar radius:

$$t_c \simeq \frac{R}{v_c} = \frac{GM^2}{LR} \frac{1}{\alpha}. \qquad (9.18)$$

This time is on the order of magnitude of the Kelvin time of the star (eq. 1.3) times $\frac{1}{\alpha}$. This expression yields a measure as to how much the meridional circulation can cause mixing of the composition. In stars with low masses, α is less than one-hundredth and the circulation time is long.

The considerations set out above, which led us to the inferred meridional flow, were made without taking into account the effects of viscosity, magnetic fields, and variations in chemical composition. A complete treatment for the configuration of a rotating star should take all these effects into account.

North pole

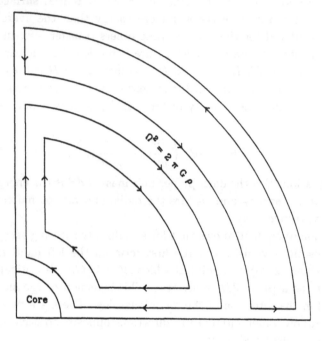

r

Figure 9.2. Meridional flow in a low mass star.

9.5   Models of Rotating Stars

It is clear that when the centrifugal acceleration resulting from the rotation overtakes the gravitational acceleration, matter will be ejected from the star at the equator. We derive a critical rotational velocity, v_{crit}, at which the matter at the equator becomes unbound, using the ratio of the centrifugal to the gravitational acceleration, $\alpha = 1$:

$$\alpha = \frac{\Omega^2 R^3}{GM} = \frac{v_{crit}^2 R}{GM} = 1 \implies v_{crit} = \sqrt{\frac{GM}{R}}. \qquad (9.19)$$

Inserting for M and R their solar values, we obtain for the critical velocity of the Sun: $v_{crit} = 440$ km sec^{-1}. The observed rotational velocity of the Sun at its equator is about 2 km sec^{-1}.

From observations we find that lower mass stars are slow rotators while higher mass stars are fast rotators. Rotational velocities of high mass stars are displayed in tab. 9.1 (v_e is the rotational velocity at the equator).

Table 9.1. Rotational velocities of high mass stars.

Spectral group	range of masses $(M_\odot)$	$v_e(\text{km sec}^{-1})$
B	5 – 18	150 – 250
A	2 – 4	50 – 150
F	1.2 – 2	10 – 50

The relation between the stellar masses and their rotational velocity is explained by their different history during their formation phase. As mentioned in Chapter 8, when the convection reaches the stellar surface it causes a strong wind. The matter ejected by the wind carries a large amount of angular momentum. The wind thus causes a decrease in the stellar angular momentum and a slowing of the stellar rotation. In addition, the stretching of the magnetic field lines brought about by convection and the differential rotation leads to a magnetic braking which further slows down the rotation.

Mestel[3] proposed a detailed model for the magnetic braking. According to this model, the crucial parameter for the efficiency of magnetic braking is the ratio of the specific kinetic energy of matter ejected in the stellar wind to the energy of the magnetic field.

The matter in the corona above the stellar photosphere is ionized, indicating that its conductivity is high. When charged particles move in a direction perpendicular to the magnetic field lines, they are forced into a circular motion around the field lines. Long range motions of charged particles are possible only *along* the field lines. Thus the matter, which is limited to moving in the direction of the magnetic field, acts as though the magnetic lines were frozen within. In a static star, the field lines may retain radial directions over a large distance from the star. In a rotating star, the situation is different. The field lines are fixed in the stellar surface where they emerge from the star while rotating with it. The energy of the magnetic field per unit volume is $\epsilon_B = B^2/8\pi$. The kinetic energy, ϵ_k, of matter ejected in the wind per unit volume is $\epsilon_k = \frac{1}{2}\rho v^2$, where v is the wind velocity. We define δ as the ratio ϵ_k/ϵ_B. The velocity of Alfven waves, v_A, is the velocity of hydromagnetic waves in an ionized matter: $v_A^2 = B^2/4\pi\rho$. Thus the ratio δ is given by:

$$\delta = \frac{\frac{1}{2}\rho v^2}{B^2/8\pi} = \frac{v^2}{v_A^2} \tag{9.20}$$

which can be defined as the square of an Alfvenic Mach number.

Close to the stellar surface and especially over the equator, δ is small. The magnetic energy prevails, and the matter is forced to corotate with the star. Far from the star, particularly over the poles, the kinetic energy of the wind prevails and the motion of the matter becomes radial. The transition between the two regions takes place on a surface located at a distance r_A from the star, at which $\delta = 1$. Up to this distance, the matter which is ejected in the wind corotates with the star. Since the radius of the rotating volume becomes very large, the rotational velocity increases and the specific angular momentum becomes very high. Thus as a consequence of the corotation enforced on the ejected matter by the magnetic field, the star loses a significant amount of angular momentum, thereby "braking" the rotation.

The specific angular momentum of a unit mass on the equatorial plane at the distance r_A is $j_0 = \omega r_A^2$, where ω is the angular velocity which is kept constant up to r_A. The specific angular momentum decreases when moving from the equator to the poles, and $j = j_0 \sin^2 \theta$, where θ is the angle measured from one of the poles. Integrating this expression over all angles and multiplying by the mass loss rate from the star, $\dot{m}$, we find that the rate of loss of angular momentum J is:

$$\frac{dJ}{dt} = -\frac{2}{3}\dot{m}j_0. \qquad (9.21)$$

This loss corresponds to a braking of the stellar rotation by magnetic effects. We obtain a time scale, τ, for the braking by dividing the total angular momentum of the star by the rate of angular momentum loss: $|\frac{dJ}{dt}| = J/\tau$.

As an example, the solar angular momentum is about 1.75×10^{48} gm cm^2 sec^{-1}, its rate of mass loss is $\dot{m} \simeq 4 \times 10^{-14} M_\odot$ yr^{-1}, and $r_A \simeq 20R_\odot$. Using these values we find that the present time scale for magnetic braking of the Sun is approximately 10^{10} yr, which is the same order of magnitude of the nuclear time scale. Higher rates of mass loss shorten this time scale significantly. For a typical B star, with mass loss rate of $10^{-9} M_\odot$ yr^{-1} and $r_A \simeq 10R_\odot$, the time scale for magnetic braking is about 10^8 yr. However B stars are high mass stars, and this time scale is on the order of their lifetime.

The creation of a magnetic field depends on the presence of a convective region in the outer part of the star. High mass stars have convective envelopes only for a short period after their formation. The period in which magnetic braking is effective in such stars is therefore short, and they retain high rotational velocities. The higher the stellar mass, the shorter is the time that the star remains in the phase of slowing rotation. Low mass stars, which retain convective envelopes throughout their main sequence phase, lose most of their angular momentum by magnetic braking and become very slow rotators. This is the case of stars of the spectral groups G and higher.

A recent investigation[4] of a group of stars in the constellation of the Pleiades showed that in low mass stars there is an inverse correlation between stellar age and the star's rotational velocity. The Pleiades are a group of young stars aged about 7×10^7 yr. The low mass stars in this group have just started their main sequence evolution (the ZAMS phase). Researchers determine the age of the stars in this investigation by the abundance of the element lithium at their surface. At high temperatures lithium is marginally stable. It burns with hydrogen through the PP II channel to form two helium nuclei at temperatures on the order of three million degrees. Thus the lithium can survive for as long as it remains at the stellar surface. If, however, the outer layers of the star are convective, then the lithium mixes inward into hotter layers, in which it burns immediately. The abundance of lithium was observed on the surface of a group of eight K stars in this constellation whose rotational velocities were measured ($v_e \sin i$). The relative abundance of lithium is determined by comparing the width of the lithium spectral lines to the parallel spectral lines of calcium (Ca), which is expected to be nearly constant in all these stars. Table 9.2 displays the correlation between the relative lithium abundance and $v_e \sin i$ of these stars. We clearly see that with decreasing lithium abundance, the rotational velocity decreases as well.

Table 9.2. Lithium abundance and rotational velocities in eight stars in Pleiades.[4]

Stellar name	$v_e \sin i$ km sec^{-1}	lithium abundance
H II 1883	> 100	1.21
H II 3163	60	0.97
H II 2244	50	1.15
H II 2034	> 60	0.03
H II 97	< 10	0.138
H II 1454	< 10	0.397
H II 879	< 10	0.419
H II 2741	10	0.430

Stars of the spectral group K definitely have a convective regime in their outer envelope that is a continuation of the convection governing the entire star before it began its ZAMS evolution. The lithium is gradually destroyed in the course of the first stage of evolution. Its decreasing abundance measures the stellar age. At the same time the stellar wind and the magnetic field "brake" the stellar rotation. This braking is effective in the presence of a significant convective region in the stellar envelope. The correlation between the decrease in lithium abundance and the decrease in rotational velocity shows that both these processes advance with time in stars with a convective region in their envelopes.

The connection between stellar rotation velocities and the age of stars suggests that rotational velocities of low mass stars may be used as a measure for their age.[5] Owing to the magnetic braking in a low mass star, rotational velocity decreases with age. The rate of this decrease is inversely proportional to the stellar mass because the convective zone, which causes the magnetic braking, is deeper in low mass stars. This suggestion can be tested by observing the rotational velocities of stars in the constellation Hyades, which is a young star cluster aged about 5×10^8 yr. We find that for stars with masses below $1.1 M_\odot$, all of which we suppose to be of the same age, the rotational velocity decreases with the mass of the star. This is true because the lower the stellar mass is, the deeper the convection zone in its envelope. The conclusion is, therefore, that in the range of ages from 10^8 to 10^9 yr, the rotational velocity of a given mass star indicates its age. At ages greater than 10^9 yr, rotational velocities become so low that this method no longer applies. Rotational velocities of stars are determined by the Doppler shifts and broadening of spectral lines, whose observation does not depend on distance. This method thus enables us to study stellar age without distance-dependent observation.

A special case is the group called Be-stars. Because of their luminosity these stars belong to the spectral group B, and the letter e in their name signifies the presence of strong emission lines in their spectra. The observed emission lines are mainly of hydrogen, which indicates that these radiative lines were produced in a hydrogen-rich matter. The stars are exceptional for having very high rotational velocities: $v_e \simeq 500$ km sec^{-1}, which is close to their critical velocity. When the centrifugal acceleration at the equator approaches the gravitational acceleration, the matter present becomes unbound. Since thermal pressure exists in the star and exerts an outward force on matter, the star begins to eject matter at the equator before the two accelerations become equal. This matter becomes a hot disk around the Be-stars, and the emission lines noted in their spectrum originate in this disk.

Some of the Be-stars reside in binary systems. Of these binaries more than 25 have been found to be pairs of Be and neutron stars. In such systems most of the mass ejected from the Be-star accretes to the neutron star, leading to bursts of X-ray radiation from the neutron star. We believe that the X-ray bursts are due to heavy mass accretion from that ejected from the Be-star. The explanation for these bursts[6] is that the system is eccentric. We consider the neutron star to be moving in an ellipse around the Be-star. When the neutron star is distant from the Be-star, it accretes mass at a low rate from the tenuous wind of the latter, and responds with a low rate of X-ray radiation. But when the neutron star approaches the periastron, it enters the matter cloud around the Be-star. High mass accretion then takes place driving X-ray luminosity up to 10^{38} erg sec^{-1}. This situation is demonstrated in fig. 9.3.

Another phenomenon involves the variations in surface brightness over the stellar surface that result from variations in the effective gravity. These are called *ellipsoidal light variations*. They originate from the variations in the effective gravity over the stellar surface due to centrifugal acceleration. As mentioned in the preceding section, the radiative flux from an equipotential surface is proportional to the effective gravity. Because of the high rotational velocity of Be-stars, their effective gravity and radiative flux almost vanish close to their equator. Since a Be-star radiates a luminosity of around $10^4 L_\odot$, the variations in the radiative flux between the poles and the equator are highly pronounced. The effective temperature at the equator may also be lower by several thousand degrees than that at the poles.

The question of how fast rotators like Be-stars form remains as yet unresolved. The highest rotational velocity that a star can have is its critical velocity. Newborn stars, which can have such velocities, must have undergone loss of angular momentum in the course of some stage in their evolution. Usually one cannot expect to find mature stars with a rotational velocity close to their critical velocity. It is commonly held that these fast rotators spun up through mass accretion in binary systems.

A Be-star in a Be neutron star binary was probably the lower mass companion in the initial binary system. The other companion, the one with higher mass, evolved faster and passed through the giant branch phases. This star completed its evolution with a supernova explosion whose remnant is the neutron star. During its red giant phase it transferred mass

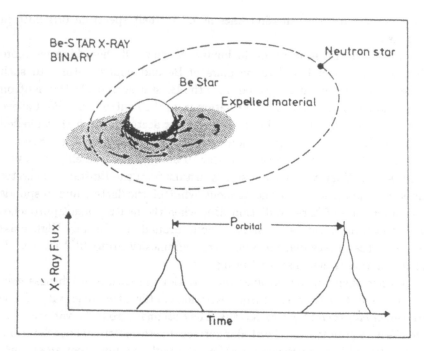

Figure 9.3. Schematic model of a Be-star X-ray binary system such as AO 538-66 and VO 332+53. The neutron star moves in a moderately eccentric orbit around the Be-star, which is much smaller than its own critical equipotential lobe. The rapidly rotating Be-star is temporarily surrounded by matter expelled in its equatorial plane. Near its periastron passage the neutron star enters this circumstellar matter and the resultant accretion produces an X-ray outburst lasting several days to weeks. [Adopted from Epstein, Lamb, and Priedhorsky.[7]]

to its companion. The transferred mass carried angular momentum, which accelerated the rotation of the accreting star and brought its rotational velocity to the limit of the critical velocity. The lower mass star became a fast rotating high mass star, or Be-star. The neutron star left by the supernova explosion is the partner of the Be-star in the binary system. The supernova explosion was also probably responsible for the eccentricity of the system.

No Be-stars have been observed in close binary systems. The reason for this is that in close binary systems their tidal effects slow down the rapid rotation of the companions. Thus even if a Be-star did exist in a close binary system, it would have become a slow rotator due to the tidal interaction.

In the future evolution[6] of Be neutron star binaries, the Be-star will go on to form a red giant, and its extended envelope will engulf the neutron star. Friction will cause the neutron star to lose kinetic energy and start spiraling inward toward the giant's core. The heating brought about by the friction will accelerate mass loss from the giant's envelope until the envelope is entirely consumed. If the remaining core of the giant is above $4M_\odot$, a supernova explosion will follow, resulting in a two neutron star binary. Examples of binary systems of two neutron stars are PSR 2303+46 and PSR 1913+16. If in the same scenario the giant completes its evolution with a lower mass core, it will not explode and its remnant will be a high mass white dwarf. The result is a binary system of a neutron star and a white dwarf. Such a system is PSR 0655+64.

9.6 The Rotating Sun

Since the Sun is the closest star, we can study its properties in greater detail than those of distant stars. It can therefore serve as a test for theories of stellar evolution. The observed properties of the Sun are its mass, radius, luminosity, and the composition at its surface. We infer the age of the Sun from indirect observations, and can actually measure its neutrino radiation.

The rotation velocity at the equator of the solar surface is seen clearly and found to be 2 km sec^{-1}. There is no way however to directly observe the rotation velocity of the inner part of the Sun. We must deduce this property by indirect methods, which are of two main types. One method is to assume a certain distribution of angular velocity in the Sun and to study the influence of this distribution on the observed properties in the present Sun. The second method is to estimate how the distribution of angular velocity would influence other stages in solar evolution, and compare this estimate with observed solarlike stars at different evolutionary stages.[8]

In studying the rotation of a star, several assumptions must be made: the amount of the initial angular momentum of the star and its distribution; the rate of loss of angular momentum from the star; and the rate of redistribution of angular momentum by transfer in the star. The last parameter is important because a star loses angular momentum from its surface. If the rate of redistribution of angular momentum is on the same order of magnitude as the rate of angular momentum loss, then the distribution of the angular momentum in the star remains in a steady-state while its total amount decreases. If the rate of redistribution is significantly lower than that of angular momentum loss, then the distribution of the angular momentum is continuously changing. It moves toward the formation of a differential rotation in the star, the inner part of which rotates faster than the outer part. We assume that a convective zone rotates as a solid body. Seismological observation of the Sun supports this assumption.

During the first stage of its evolution along the Hayashi track, a star is fully convective and rotates as a solid body. This assumption determines the distribution of the angular momentum. The single parameter needed as an initial condition is the amount of the total angular momentum, J_0. Massive stars become radiative immediately after their Hayashi track phase terminates. Since significant loss of angular momentum from a star depends on the existence of an outer convective zone, we can assume that massive stars, whose envelopes are radiative, did not lose a significant amount of angular momentum at this phase. We can assume that they possess about the same total angular momentum today as they did at ZAMS (Zero Age Main Sequence). The rotation velocities of massive stars today are high and are some fraction of the critical velocities (see tab. 9.1). We may assume that the rotation velocities of all stars are the same fraction of their critical velocities at ZAMS. This reasoning leads to the assumption that the angular momentum of the Sun at ZAMS, is $J_0 = \alpha \times 1.63 \times 10^{50}$ gm cm^2 sec^{-1}, where α is a constant on the order of unity.

The matter that leaves the star by a stellar wind carries angular momentum. If this matter is locked to the stellar rotation by the magnetic field lines (as was explained in a preceding section), the amount of lost angular momentum is significant. This happens in stars having outer convective zones. While the outer part loses angular momentum and its rotation velocity decreases, the stellar core contracts because of the increase in average atomic weight due the nuclear transmutation. The conservation of angular momentum demands that during the core contraction, the decrease in the core radius leads to an increase in the rotation velocity. A differential rotation thus forms in which the envelope rotates more slowly than the core. At the boundary layer between the two regions, a torque forms which drives a flow of mass and angular momentum. This flow causes mixing of the composition in a broader region than ordinarily expected from mixing by convection alone. This observation of greater mixing than expected from convection shows that rotation mixing exists.

We referred earlier to the lithium problem. It is assumed that the lithium abundance found in meteorites represents that at the formation epoch of the Sun. Comparing the amount of lithium present at the solar surface today to that found in meteorites shows a depletion by a factor of 200 during the lifetime of the Sun. As noted above, lithium is destroyed at a temperature of about three million degrees and higher. Calculation of the standard model of the Sun yields a temperature of two million degrees at the base of the convection zone. This temperature is too low to be the cause of the lithium depletion observed in the Sun, which demands mixing into deeper and hotter layers. This extra mixing can be caused by rotation. Thus the present abundance of lithium serves as a probe for studying the rotation of the inner part of the Sun.

Beryllium is another element whose surface abundance depends on inward mixing. This element is destroyed at a temperature higher than that needed for the lithium burning. The depletion of beryllium, therefore, demands deeper mixing than is needed for lithium destruction. The depletion of beryllium at the solar surface is by a factor of three. This also indicates the extent of mixing below the convection base, where rotation causes the extra mixing.

We compare the present abundance of these elements with their abundance at the formation epoch of the Sun as inferred from the amount of elements found in meteorites. However we should note that this abundance is higher than the primordial abundance as inferred from cosmological theories of big bang nucleosynthesis (BBN). The proposal used to explain this difference is that during the lifetime of the Galaxy, interactions between cosmic rays and elements in the interstellar medium enriched the abundance of the light elements.[9]

Another probe for the inner structure of the Sun may be inferred from observations of oscillations in the Sun. Oscillations on the solar surface have periods of about five minutes. Oscillations form inside the Sun. By observing them at the surface, we can study the properties of the inner parts. There are two types of stellar oscillations, called p-mode oscillations and g-mode oscillations. In p-mode oscillations the restoring force results from pressure, whereas in g-mode oscillations the restoring force results from gravity. From analysis of the oscillation theory, we find that the main contributions to p-mode oscillations come from the outer part of the star. However, the main contributions to g-mode oscillations come from the inner tenth part of the star and form in nonconvective regions. Thus observations of g-mode oscillations would yield the most important information for studying the inner part of the Sun. Unfortunately, the main observations of solar oscillations to date contain p-mode oscillations. We must await improvements in the observations of solar oscillations to obtain the vital information that might be gained by investigating g-mode oscillations.

The degree of the difference in the rotation velocity depends on the efficiency of the transfer of angular momentum from the central part to the surface. Here the angular momentum is lost from the star through stellar wind. From observations of solarlike stars, we find that the characteristic time for such a star to reduce its surface rotation velocity to the order of a few km sec^{-1} is a few times 10^8 yr. This time scale does not depend on the amount of the initial angular momentum because rapidly rotating stars will produce stronger magnetic braking than slowly rotating stars. Thus after few times 10^8 yr, the star "loses memory" of its initial rotation velocity.

To estimate what happens to the rotation velocity of the inner parts, we may study the properties of solarlike stars at later stages in their evolution. We mentioned in Chapter 6 that upon completing its main sequence evolution, a star turns into a red giant with a dense core and an extended, mostly convective envelope. In such a configuration, the core remains decoupled from the envelope, and may retain its original rotation velocity. During its evolution along the red giant branch (RGB), the star loses mass through a significant stellar wind. It may lose from 0.15 to 0.2 $M_\odot$ in this phase. The red giant phase terminates with the ignition of the helium at the core, when the star begins its evolution along the *helium main sequence* phase. This period is called the *horizontal branch* in the H-R diagram. At this stage the stellar core expands to a greater extent than it was on the RGB, the stellar envelope contracts, and the convective zone narrows and practically vanishes. The core and the envelope couple more strongly than they did on the RGB. If the core has a higher angular velocity than the envelope it will accelerate the envelope, transferring angular momentum from itself to the envelope.

Observations of solarlike stars in their horizontal branch phase show rotation velocities of 30 km sec^{-1}. Such a high rotation velocity can be achieved by a single star only if it had some unobserved "store" of angular momentum while it was a main sequence star. Such a store may be a fast rotating core in the main sequence phase of the star. This observation leads us to conclude that a differential rotation should exist in a solarlike star in its main sequence phase.

We observe rotation velocities of 20 km sec^{-1} for several white dwarfs. A solarlike star which rotates as a solid body at its main sequence phase with rotation velocity of 2 km sec^{-1} like our Sun will produce a white dwarf star with a maximal rotation velocity of 5 km sec^{-1}. This observation also leads us to conclude that a solarlike star in the main sequence phase should have a configuration with a differential rotation.

Summing up all of the above considerations, we can conclude the following about the model of the present Sun[8]:

(1) The outer part of the Sun ($r > 0.6R_\odot$) rotates slowly and with almost uniform angular velocity.

(2) The inner part of the Sun ($r < 0.4R_\odot$) forms a rapidly rotating core. In the star formation phase, the Sun had the same angular momentum as the progenitor cloud. During evolution toward the main sequence, the proto-Sun lost angular momentum from the outer mass layers, and this part of the Sun rotated slowly.

The existence of differential rotation may explain several curious features in the surface abundance of elements in the present Sun. These features can be explained if we relate them to mixing by meridional and radial flows induced by rotation.

References

1. Tassoul J.L., 1978, *Theory of Rotating Stars*, Princeton University Press, Princeton.

2. Kovetz A., Shaviv G., Zisman S., 1976, *Astrophys. J.*, **206**, 809.

3. Mestel L., 1968, *Mon. Not. Roy. Astron. Soc.*, **138**, 359.

4. Butler R.P., Cohen R.D., Duncan D.K., Marcy G.W., 1987, *Astrophys. J. Lett.*, **319**, L19.

5. Kawaler S.D., 1989, *Astrophys. J. Lett.*, **343**, L65.

6. Van den Heuvel E.P.J., Rappaport S.A., 1988, in *Physics of Be-Stars*, ed. A. Slettebak, T.D. Snow, Cambridge University Press, Cambridge.

7. Epstein R.I., Lamb F.K., Priedhorsky W.C., 1986, *Los Alamos Science*, **13**, ??.

8. Pinsonneault M.H., Kawaler S.D., Sofia S., Demarque P., 1989, *Astrophys. J.*, **338**, 424.

9. Deliyannis C.P., Pinsonneault M.H., 1990, *Astrophys. J. Lett.*, **365**, L67.

Chapter 10

Supernova

Supernova explosions are very prominent phenomena which are already found recorded in ancient chronicles. The name *nova* derives from the Latin word meaning "new" and refers to the sudden appearance of a star where it was not observed earlier. This "new" appearance is due to the event of a nova explosion, during which the luminosity of the star increases by many orders of magnitude. It becomes visible over very large distances, even when it occurs in other galaxies.

Researchers found that in a group of nova stars there is a subgroup whose energy release is greater by a few orders of magnitude than the average energy ordinarily released in nova events. This subgroup was given the name *supernova* (SN). The amount of energy released in an SN event in the form of radiation and kinetic energy of matter is around 10^{51} erg. Such an event produces a significant amount of heavy elements and releases them into the interstellar medium. To date researchers have found and catalogued more than 600 extragalactic SN events.

The SN group divides into two subgroups: SN Type I (SN I) and SN Type II (SN II). The reason for this division is the observed presence of strong hydrogen emission lines in the spectrum of SN II and the absence of such lines in the spectrum of SN I. This shows that the progenitors of SN II possessed a hydrogen-rich envelope before the explosion. The two SN types are entirely different both in their character and in the evolutionary track which leads to the SN explosion.

SN II form from massive stars ($M > 8M_\odot$) which reach the stage of producing iron in their cores by nuclear reactions. The SN event occurs as a result of the gravitational collapse of the iron core. Usually in such an event a neutron star forms which may later appear as a pulsar. SN I form from white dwarfs located in close binary systems and whose mass increases above a critical limit because of mass accretion from their companions. This increase causes an explosive ignition of nuclear reactions of heavy elements in the stars' cores. Ordinarily no neutron star forms in such an event. In an SN II event, the total energy released in the collapse is on the order of 10^{53} erg, of which the main part radiates as neutrinos. A few times 10^{51} erg converts to the observed luminosity and kinetic energy of the ejected matter. In an SN I event, the released energy is on the order of 10^{51} erg, most of which converts to observed luminosity and kinetic energy of the ejected matter. For this reason the two types of events have similar observational features.

SN II distribution follows the distribution of population I stars. They appear in the arms of spiral galaxies. SN I distribution follows that of population II stars and appear in all the galaxies. According to statistics accepted today, an SN forms every 50 years in our Galaxy, with an almost equal weight of types I and II.

Bear in mind that by direct observation we can study the behaviour of the outer envelope and the luminosity of the star. The processes occurring inside the star responsible for the behaviour of the envelope take place under the cover of the outer part of the star. We may infer the characteristics of these processes from their results which appear in the behaviour of the envelope.

The first historically known supernovae[1] were recorded in Chinese historical records. European records mention only that of 1006 and the stars of Tycho Braha and Kepler. The following, then, are recorded supernovae in their historical order of appearance:

SN 185 exploded on 7 December 185 AD, and was observed for almost two years between the stars α and β of the constellation Centaurus. Its remnants are seen mainly in the radio wavelengths as RCW 86.

SN 386 exploded in 386 AD in the constellation Sagittarius, and was observed for three months. Its remnants are identified as G11.2-0.3.

SN 393 exploded in the constellation Scorpio, and was observed for eight months.

SN 1006 exploded close to the constellation Scorpio on 1 May 1006. It was seen for several years. It was widely reported in Europe, and its remnants are identified as PKS 1459-41.

SN 1054 exploded in the constellation Taurus on 4 July 1054. It was noted for more than one year, and its remnants are identified as the Crab Nebula.

SN 1181 exploded in the constellation Casiopeia, and was observed for half a year. Its remnants are identified as 3C 58.

SN 1572, known as Tycho's star, exploded in Casiopeia on 6 November 1572. It was seen for eighteen months.

SN 1604 is known as Kepler's star. It exploded between the constellations Ophiuchus and Sagittarius on 10 October 1604. It was observed for a year.

A remnant in Casiopeia constellation, named Cas A, was left by an SN. Probably this is a remnant of Flamsteed's star,[2] observed in 1670.

10.1 Supernova Type II

The lower limit of initial stellar masses that can lead to the formation of an SN II is about $8M_\odot$. This is the upper limit of initial stellar masses that lead to the formation of white dwarfs. (The URCA process, which is discussed in Chapter 12, may push this limit to even lower values). Stars with initial masses above this limit ignite carbon in their cores while the matter is not in a degenerate state. As a result, the ignition starts relatively calmly. Stars with initial masses of above $11M_\odot$ pass gradually through all the evolutionary phases of producing carbon, oxygen, neon, silicon, up to iron in their cores, without developing degeneracy. Stars with initial masses in the range 8 to 11 $M_\odot$ form limit cases in which partial degeneracy develops before producing iron. The explosion which follows the collapse may be more complicated.

The upper limit on initial masses of SN progenitors should be that mass which allows a hydrogen-rich envelope to exist when the stellar core collapses. This limit is at around $40M_\odot$. There are higher mass stars, but the hydrogen-rich envelope of these is consumed before the SN explosion. Therefore no hydrogen lines appear in their spectrum.

The energy production per unit mass by nuclear reactions is very low in the burning of heavy elements and is less than one-tenth of energy release in hydrogen burning. Because of the high temperatures at which these phases take place, energy loss by neutrinos during the burning is very high. For these two reasons, the burning phases of heavy elements are very short. Usually when the core evolution goes through the phases of burning the heavy elements, the envelope is in the expanded form of a red giant.

Stars with initial masses of 10 to 11 $M_\odot$ may eject their hydrogen-rich envelope a few years prior to the SN explosion. When the following explosion shock overtakes the expanding envelope, a very high luminosity forms. This phenomenon probably occurs in about one percent of SN II. We think that the SN which created the Crab Nebula, exploding in the constellation Taurus in the eleventh century, underwent an event of this nature. The event is characterized by low percentage of heavy element lines in the spectrum.

When the iron core becomes so massive that the pressure created by the (partially) degenerate electrons cannot support it against gravity, the core collapses. The processes which lead to the explosion of stars whose masses are of 15 to 20 $M_\odot$ and those involved in the explosion of higher mass stars are different. In the case of the former, there are two processes which accelerate collapse: (1) the capture of electrons by iron group elements reduces pressure, which results from the decrease in the number density of free electrons; (2) the neutrinos released during this capture radiate outward and further reduce the energy content of the core.

In stars with masses above $20M_\odot$, the pressure falls because of photodisintegration of the iron atoms. At very high temperatures, the iron atoms disintegrate to α particles due to the absorption of high energy photons created in the high temperature. The rapid contraction (collapse) releases a great deal of of gravitational energy, most of which is absorbed in the disintegration of the iron atoms. The core continues to collapse, and the matter around it falls at supersonic velocity — almost a free-fall. Part of the gravitational energy converts to heat and kinetic energy. This energy is later observed as the stellar luminosity and kinetic energy of the ejected matter. The important link now is to understand how this energy (or part of it) couples with the envelope and causes it to expand in an explosive manner. The common view today is that this coupling forms through a transfer of mechanical energy by a hydrodynamic shock wave.

The relevant physics for calculating the collapse of the core consists of nuclear physics (especially the rates of weak interactions through which neutrinos form), the equation of state for high density matter, and the cross-section for the interaction of the neutrinos with matter. The scenario which emerges is the following:[3,4]

Prior to the explosion relativistic electrons create most of the pressure. When the density reaches the value of 10^{11} gm cm^{-3}, the cross-section for interactions of neutrinos with matter becomes so high that the neutrinos are trapped in the core. (This means that their diffusion time outward is longer than the time scale of the collapse, which is few milliseconds.) Because of the high heat capacity of the heavy nuclei, the core heats up slowly and continues to contract up to the nuclear density ρ_{nuc}, which is about 2.8×10^{14} gm cm^{-3}. The characteristic mass scale of the region that reaches this high density is 0.6 to 0.8 $M_\odot$, which is about half the iron core.

When the density becomes higher than ρ_{nuc}, the equation of state becomes stiff (that is, the nuclear force becomes repulsive instead of attractive, and the pressure becomes infinitely high). The collapse of the inner part of the iron core suddenly halts while the outer part continues infalling at a speed of about 70,000 km sec^{-1}. At this moment the density reaches values of $\simeq 3\ \rho_{nuc} \simeq 8 \times 10^{14}$ gm cm^{-3}, and the temperature reaches a value of 10^{11} degrees. The total pressure now comes from the relativistic electrons, nuclear pressure, and additional pressure created when the nuclei of the atoms transform into a fluid of free nucleons.

When the contraction stops, pressure waves form and begin traveling outward with a velocity equal to the velocity of sound. These pressure waves accumulate and form a shock front which moves outward. The shock wave does not form at the centre but outside the central superdense core around a mass sphere of 0.8 to 0.9 $M_\odot$. The gravitational energy released in the collapse is on the order of 2 to 3 $\times 10^{53}$ erg. Ninety-eight to 99 percent of this energy is released in the form of neutrinos, and few times 10^{51} is stored in the shock wave. The shock wave moves outward through the outer part of the core which continues to fall inward. It loses a significant part of its energy during this phase, especially by neutrino emission, because the lower density matter becomes transparent to neutrinos. The temperature in the shock wave is so high that the iron nuclei in the matter break up into single nucleons. This process absorbs energy at the rate of 1.7×10^{51} erg per each $0.1 M_\odot$ of iron converted to single nucleons.

Two zones are distinguishable in the iron core: an inner superdense core, around which the shock wave forms; and an outer lower density zone, through which the shock wave travels, breaking up the iron nuclei into nucleons. For the SN eruption to be maintained, the shock wave must reach the outer edge of the iron core before losing all of its energy. When the shock wave reaches a lower density zone its temperature decreases, becoming too low for the iron decomposition and less efficient in neutrino production. Hence to sustain the eruption, the iron core mass must be sufficiently low so that the zone between the location of where the shock forms and the edge of the core cannot absorb all the energy of the shock wave. The upper limit of an iron core mass enabling an SN explosion is likely about 1.25 to 1.35 $M_\odot$. The time scale for the evolution of the scenario described above is about 20 milliseconds.

According to this scenario, iron cores with masses higher than $1.35 M_\odot$ will not lead to an SN explosion and will not create neutron stars. In such stars the shock dies out before leaving the iron core and the whole system collapses, probably to form a black hole.

Wilson[5] proposed another scenario of the evolution of an SN explosion which is supported by his own numerical simulations. This scenario is called the *delayed explosion mechanism.* It assumes that when the shock wave in the iron core stops, it will gain further energy from the neutrinos, which diffuse slowly behind the shock. A few percent of the neutrinos produced in the core's collapse are absorbed by the neutrons and protons behind the shock. Their energy transforms into heat energy which replenishes the shock. The repumped shock wave proceeds outward at lower speed than initially but is still able to reach the envelope with sufficient energy to eject the envelope explosively. Bethe[6] showed that a very efficient convection that develops behind the shock wave helps in repumping the shock. The process of stopping and repumping the shock wave can repeat several times. The advancing shock turns into an explosion on reaching the oxygen burning shell, where the density is significantly lower. Such a scenario can lead to an SN explosion even in stars with masses of 20 to 40 $M_\odot$. We estimate that the energy released in the form of kinetic energy and stellar luminosity is lower in the case of the delayed explosion mechanism than it is in that of the hydrodynamic mechanism.

10.2 Explosive Nuclear Reactions

When the shock wave moves outward it leaves behind a very hot zone. We estimate the temperature of this zone from an equation for the energy content, E_0, of the spherical volume left behind the shock wave:

$$E_0 = \frac{4\pi}{3} R^3 a T^4 \tag{10.1}$$

where $E_0 \simeq 10^{51}$ erg, R is the radius of the sphere enclosed by the shock wave, and T is the average temperature of the zone. Hence:

$$T = \left(\frac{3E_0}{4\pi R^3 a} \right)^{1/4}. \tag{10.2}$$

We note that the temperature decreases with the advancement of the shock.

When the temperature is above 5×10^9 degrees, any given composition will relax to nuclear equilibrium on a time scale of less than one second. The matter, enclosed in a radius of 3,500 kilometres, reaches this temperature threshold, and mainly transmutes to elements of the iron group. Most of this matter turns into ^{56}Ni. The amount of ^{56}Ni produced (which is later transformed into ^{56}Fe) depends on the radius of the hot sphere. If the density gradient in the core is steep, then the radius is small. Low mass stars, with masses of 8 to 11 $M_\odot$, have partially degenerate cores with a steep density gradient. They produce about $0.003M_\odot$ of ^{56}Ni.

When the shock wave moves outward its temperature decreases, falling below 2×10^9 degrees around the oxygen-neon shell, and the explosive nuclear reactions die out. The explosion produces elements heavier than magnesium. Lighter elements are products of regular evolution that took place before the explosion.

After about a minute the shock wave reaches the outer edge of the helium shell, which is at a radius of 5×10^{10} cm. If the star lacks a hydrogen envelope, the helium is ejected with velocities of 5,000 to 25,000 km sec^{-1}. This is probably the case of SN Ib (see the sections below). In the presence of a hydrogen-rich envelope, the motion of the matter brakes, and a braking wave moves inward.

The original shock wave travels in the hydrogen envelope with the velocity given by:

$$R_S = \left(\frac{E_0}{\rho}\right)^{1/5} t^{2/5} \qquad (10.3)$$

where R_S is the radius of the shock. Assuming $\rho = 3M_{env}/4\pi R^3$, where M_{env} is the envelope mass and the stellar radius equals 3.5×10^{12} cm, we find for t_b, the time needed for the shock to reach the stellar surface:

$$t_b \simeq 2500 \left(\frac{M_{env}/M.}{E_0}\right)^{1/2} \text{sec.} \qquad (10.4)$$

10.3 The Light Curve of an SN II

From observations we are able to study what happens in the envelope as a result of the explosion taking place in the core. The factors which determine what will be seen are the following: the amount of energy deposited in the envelope; the composition and the density distribution in the envelope before the explosion; and the processes of energy production by nuclear reactions in the envelope, when it starts to expand under the impact of the exploding core's violent push.

The order of magnitude of the core radius is of 10^8 cm, while that of the envelope is of 10^{12} to 10^{13} cm. Thus in dealing with the evolution of the envelope, we can treat the core as a point mass and an energy source. Most of the observations have been made within the optical range of wavelengths.

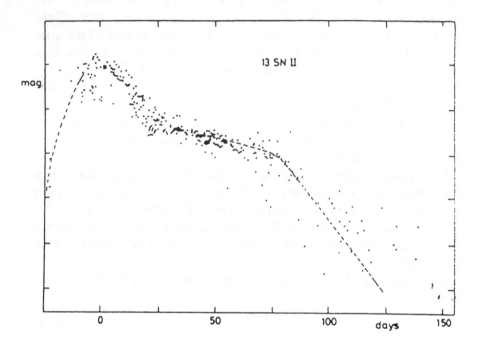

Figure 10.1. Composite light curve obtained from 13 SN II. [After Barbon, Ciatti, and Rosino.[7]]

The general trend of the light curve in most SN II is of a rapid rise to a maximum, a plateau lasting two to three months, followed by a gradual decline of the luminosity lasting for a few hundred days.

The first observed feature concerning the explosion is a sudden rise in the luminosity of up to $10^{10} L_\odot$. The envelope starts expanding, cools down, and the stellar luminosity decreases by a factor of 10. The envelope cools gradually, starting at the surface and advancing inward. Prior to this expansion the envelope is opaque. The envelope matter absorbs 99 percent of the energy of the explosion and converts it to kinetic energy. One percent of the energy, or about 10^{49} erg, radiates when the envelope matter becomes transparent. This process proceeds gradually, starting from the stellar surface. During the first stage of expansion, in which the outer radius increases by a factor of 30, the matter cools down, the ionized hydrogen recombines, and opacity falls by a few orders of magnitude. This opacity decrease is what causes the energy radiation mentioned earlier. The cooling front, followed by the decrease in opacity, advances inward and adds new layers which are now able to radiate their thermal energy. As long as the cooling front travels inward, the luminosity remains constant. The cooling front crosses the envelope in about two months and is the cause of the plateau in the luminosity curve in this period.

In the course of this phase, the effective temperature remains almost constant, at around 6,000 degrees, which is the recombination temperature for hydrogen at a density of 10^{-13} gm cm^{-3}. The outer radius of the photosphere remains constant as well (at 1.5×10^{15} cm) because the contraction of the photosphere caused by hydrogen recombination compensates for the expansion of the envelope.

The plateau in the luminosity continues as long as the cooling recombination front advances inward because the new transparent layers formed by recombination behave as a new source for radiated energy. When the recombination front reaches the core, this energy source is consumed.

Later, the luminosity forms by nuclear decay of ^{56}Ni to ^{56}Co, and the decay of ^{56}Co to ^{56}Fe. The time scale of spontaneous decay of these isotopes determines the slope of the luminosity curve. The half-life of ^{56}Ni is 6.1 days, and that of ^{56}Co is 77.1 days; thus the slope depends mainly on the decay rate of ^{56}Co. The amount of ^{56}Ni production demanded for this process is 0.1 to 0.4 $M_\odot$, and it is reasonable that the explosion of the core produces this amount. On top of the optical luminosity, there is also radiation in shorter wavelengths, in the form of UV and γ radiation.

The interaction of expanding matter with any form of matter already residing in the vicinity of the exploding star may cause additional effects. These effects are added to the basic features of the SN luminosity curve. Such a case may be an interaction with the part of the envelope ejected by some mechanism shortly prior to the explosion.

On the other hand, heavy stars close to the upper mass limit for SN creation might have very low mass hydrogen-rich envelopes. In this case, the plateau in the luminosity curve may be very short, or entirely absent, and the decay rate of ^{56}Co determines the characteristic slope in the luminosity curve.

A significant part of the formation of the heavy elements in the Galaxy takes place in SN I and SN II events. SN II usually leaves behind a neutron star formed from the iron core of the progenitor. The contribution of SN I to the Galaxy's abundance of heavy elements is therefore larger than that of SN II.

10.4 SN 1987A

On 23 February 1987, astronomers observed a supernova explosion in the
Large Magellanic Cloud (LMC), whose distance from us is about 165,000 LY.
This event immediately attracted attention all over the world. The object
was the focus of all available observation facilities. What ensued was followed very closely and presented an opportunity of comparing theory with
observation. This explosion is of a type II supernova. The explosion obviously occurred 165,000 years ago. We only detect it now because of the
time needed for the signals to reach the Earth. All the times are registered
according to the Earth clock (UT). Observers who happen to be located
elsewhere in the universe will register the event according to their clocks.

The LMC is a well-studied region. The exploded star was identified as
SK-69 202, a blue supergiant star that has been known for almost a century.
Its effective temperature was almost 16,000 degrees, and its luminosity was
about $10^5 L_\odot$. Estimates of its mass were approximately of $16 M_\odot$, of which
$10 M_\odot$ was in the hydrogen-rich envelope. Its radius was $\simeq 3 \times 10^{12}$ cm
$\simeq 43 R_\odot$. Most of the observed features accorded well with the theoretical
predictions,[4] but there were also several surprises that demanded careful
analysis by scientists.

The first surprise was the fact that the exploding star was a blue giant,
rather than the red giant expected from a theoretical evolutionary track
leading to an SN explosion. We believe that the main reason for SK-69 202
being a blue instead of a red giant is its low metallicity. In the region
where this star resides, metallicity is lower by a factor of more than three as
compared to that of the Sun, which indicates a presence of about 0.5 percent
heavy elements in the composition. The low metallicity affects the stellar
structure in two ways — first by decreasing the opacity of the hydrogen-rich
envelope, and second by reducing the efficiency of the CNO cycle in the
burning of hydrogen. Both these factors lead to the formation of a hotter
star with higher effective temperature than in a star with the same mass
and higher metallicity.

Evolutionary calculations of low metallicity giant stars had indeed produced blue giant stellar models prior to SN explosion. These evolutionary
sequence simulations also showed that the star had been a red giant until
about 40,000 years before the explosion. Since red giants lose mass by stellar wind at a higher rate than do blue giants, traces of this mass should be
found in the circumstellar matter around the star at an appropriate distance (about one light year). The interaction of this matter with the first

burst of UV radiation from the explosion was observed as two concentric rings in March through April 1988. These rings are called *light echoes*. The matter that reflects these echoes resides around the star. The time delay between the original burst and the appearance of the light echoes is due to the time needed for light to reach this location before it is reflected in our direction. When the SN ejecta reach this matter, we expect a violent interaction to occur. Since ejecta velocity is about 0.1 of light velocity, we anticipate this interaction to take place in 10 years from the explosion. A picture of the inner ring of the two light echoes is shown on the cover of this book. It was taken by the Hubble Space Telescope in August 1990. The luminosity of the ring reached its maximum about 400 days after the SN explosion, and it became weaker later.

Unfortunately the explosion of SK-69 202 did not provide us with sufficient information to distinguish between the two alternative scenarios for the explosion: the hydrodynamic scenario and the delayed explosion scenario (see Section 10.1). The energy released in a hydrodynamic explosion is above 10^{51} erg, whereas a delayed explosion releases less than 10^{51} erg. The energy released by SK-69 202 in the explosion as kinetic energy and stellar luminosity is estimated as 0.6 to 1.5 $\times 10^{51}$ erg. However, researchers made quite accurate observations of the neutrino flux produced by the explosion, of the energy spectrum and the timing of the neutrinos, and of the amount of ^{56}Ni produced in the explosion.

The neutrinos produced in the central part of the explosion reach the stellar surface very quickly and are apparent immediately. The shock wave, which begins traveling outward almost at the same time, reaches the surface later. The time delay between the observation of the neutrino burst and the appearance of the shock by a burst of luminosity reflects the time required for the shock to travel the distance from the centre to the surface. The time delay between the detection of the neutrinos and the first optical observation and the star's known radius, accord with the supersonic transference of the shock. The luminosity curve from the second month onward exactly fits the light curve expected from the decay of ^{56}Co, ejected from the core to the envelope by a strong shock wave.

In the neutrino burst we should distinguish between the neutrinos expected from the dynamic phase, which lasts for milliseconds, and those expected from the cooling phase, which lasts for tens of seconds. The flux of neutrinos created in the dynamic phase that reaches the Earth is about 10^{10} cm^{-2} sec^{-1}. It takes them a few seconds to reach the stellar surface.

The Kamiokanda II telescope in Japan and the IMB telescope in the US performed the principal neutrino observations. In two other neutrino tele-scopes — the Baxan in the Caucasus mountains and the Liquid Scintillator Detector (LSD) under Mont Blanc — the expected flux was at the limit of their observational threshold. The Kamiokanda II and IMB telescopes used purified water as both the target and detector. These telescopes are supposed to detect either electrons that rebound due to collisons with neu-trinos, according to the process $\nu + e \rightarrow \nu' + e'$ (where primes denote the scattered particles), or positrons created in the absorption of an antineu-trino by a free proton, according to the process $\bar{\nu}_e + p \rightarrow n + e^+$.

The rebounded electrons and the positrons created in the neutrino ab-sorption travel in the water with velocities higher than the velocity of light in water. In such a case a special radiation is observed, known as the Cherenkov effect. Photomultipliers arranged around the water tank record this radiation. The advantage of this type of detector is that users can calculate the energy and the direction of each event. Both telescopes are in the Northern Hemisphere of the Earth. Since LMC is in the South-ern Hemisphere, the neutrinos from SK-69 202 traversed the interior of the Earth before entering the detectors from below.

Kamiokanda II is located in the Kamioka mine in Japan and is 1,000 metres underground. The 3,000 cubic metres of the detector water are within a cylindrical tank with a diameter of 15.6 metres and a height of 16 metres. IMB is located in a salt mine under Lake Erie, near Fairport, Ohio, at the depth of 580 metres. It contains 6,800 cubic metres of water in a rectangular tank. The energy threshold for the neutrinos possibly observed by these detectors is that of neutrinos created at a temperature of above 3.7×10^{10} degrees.

Twelve neutrinos were registered in Kamiokanda II and eight in IMB. The neutrinos registered in both detectors at the same time, 23 February, 7:36. The first optical observation that corresponds to the breakout of the shock wave through the stellar surface occurred three hours after the arrival of the neutrinos. This time interval well suits that required for the shock wave to travel from the centre to the stellar surface with an average velocity of 2,800 km sec^{-1}.

A calculation of the amount of energy carried by the neutrino flux reach-ing the detectors yields 6×10^{52} erg.

The neutrinos produced in this event furnished an important test for some of the properties of these particles: their mass, their electric charge, and their lifetime.

The mass of a neutrino. If neutrinos possess a rest mass, then the time needed for their arrival will be different for neutrinos with different energies. For massive particles, the higher the energy of the particle is, the higher its velocity will be. The great distance from the source would make the spread in the arrival time more significant. From calculating the energies and arrival times of the neutrinos, we find that those observed from SN 1987A put an upper limit of 11 eV on a neutrino's mass. This is equivalent to 2×10^{-5} of the electron mass.

The electric charge of the neutrino. In a similar way, a suggested charge of the neutrino can be tested. If the neutrinos possess a charge, they will be deflected by the (known) magnetic field of the Galaxy. The deflection will be inversely proportional to the neutrinos' energies, and can be detected in the larger orbits and delayed arrival of lower energy neutrinos. The calculations yield an upper limit to the neutrino's charge of 10^{-17} of the electron charge.

The lifetime of the neutrinos. Plainly, the detection of neutrinos originating from a source located at a distance of 165,000 LY demands that they have a longer lifetime than the time needed to travel such a distance. Such a long lifetime contradicts the assumptions of a much shorter existence used to explain the deficit in the neutrino flux expected from the Sun (see Chapter 12).

10.5 Supernova Type I

The commonly accepted scenario for the formation of an SN I is of mass accretion onto a white dwarf that drives it above some critical mass. Since such a star has no hydrogen-rich envelope, the newly synthesized elements of the explosion can be observed in the spectrum.

Consider a CO white dwarf of around one solar mass. The stellar matter is in a state of very high degeneracy. The matter accreted to the star consists of mainly light elements, namely of hydrogen and helium. If the mass accretion rate is lower than $10^{-8} M_\odot$ yr^{-1}, the accrued matter remains cool and becomes degenerate. After a certain amount of matter accretes, an explosive ignition of the hydrogenic matter causes a nova explosion (see Chapter 11). When the rate of mass accretion is higher than $10^{-8} M_\odot$ yr^{-1}, the accretion creates a high temperature. This temperature causes the immediate burning of light elements to carbon and oxygen which are added to the CO main body of the star. The increase in the mass by accretion leads to an increase of the temperature in the star's inner portion, until the temperature rises above the threshold temperature for carbon ignition.

Because degeneracy is high, a pressure increase does not follow the burning, and the regulating mechanism described in Chapter 4 does not work. The nuclear burning develops into a runaway, and the huge amounts of released energy disrupt the star, ejecting the matter at very high velocities. If the burning begins off centre, a small part of the star may remain as a remnant. Usually the entire star is blown off and nothing remains. This event does not create a neutron star or a black hole.

The nuclear burning advances very rapidly through the whole chain of elements, up to the production of elements of the iron group, especially ^{56}Ni, whose half-life is 6.1 days. ^{56}Ni decays to ^{56}Co, whose half-life is 77.1 days. The decay of these elements determines the features of the light curve of the SN I. The total nuclear energy released is 10^{51} erg. As is the case with SN II, most of this energy converts into kinetic energy of the ejected matter. When the star expands from a radius of 10^9 cm to that of 10^{15} cm, it becomes transparent and begins to radiate its thermal energy. Since the white dwarf had no hydrogen-rich envelope, its luminosity contains no hydrogen lines.

The nuclear burning in the star may advance as a detonation in which the speed of the advancing burning front is supersonic, and the density and

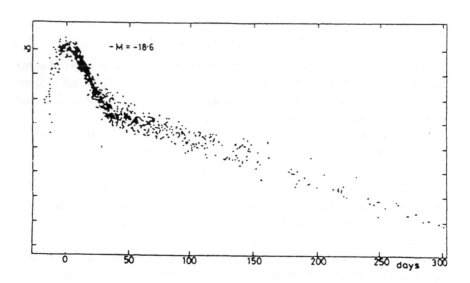

Figure 10.2. Composite light curve obtained from 38 SN I. [Adopted from Barbon, Ciatti, and Rosino.[8]]

pressure increase behind the burning front. Alternatively, it may proceed as a deflagration (or a flame) in which the burning front advances at a subsonic speed, and density and pressure decrease behind the burning front.

The standard model for an SN I is as follows:[3]

When the stellar mass increases by mass accretion, the star contracts, and the contraction releases gravitational energy and produces nuclear energy at a low rate, but the stellar centre does not heat up. As a result of energy loss by neutrino radiation, the central part is cooler and a temperature inversion forms. When the stellar mass approaches the value of $1.4M_\odot$, the central density reaches the value of 2×10^9 gm cm^{-3}. At this density the energy loss by neutrinos becomes slower; the energy production by nuclear reactions overcomes the energy loss by the neutrinos, and heating of the centre becomes efficient. The contraction continues and the density increases further to 4×10^9 gm cm^{-3}. When the temperature reaches the value of 3.5×10^8 degrees, the nuclear reactions accelerate significantly.

Neutrinos cool the most central part (a sphere of $0.02M_\odot$) through the URCA process. (The temperature threshold for the URCA process is 3.9×10^8 degrees; Chapter 12 explains the URCA process in detail.) The main burning takes place over the surface of the central sphere. The zone around the central core becomes convective. However, the convection is not efficient enough to transfer the amount of energy produced by nuclear burning, and the burning zone advances both inward and outward. The ignition of the central sphere proceeds inward from all directions, and the centre ignites in a detonation.

The burning advances as a deflagration outward — either because the ignition did not lead to detonation; or, if the process starts with detonation, the detonation decays to a deflagration which advances outward with a subsonic velocity (around 50 km sec^{-1}). A shock wave moves ahead of the burning front and pushes the matter outward. Part of the matter is ejected before it goes through further nuclear reactions. As long as the temperature at the deflagration front is above 5×10^9 degrees, most of the matter converts to elements of the iron group. It reaches this limit in a sphere of mass of $0.7M_\odot$. In the range of 0.7 to 1 $M_\odot$, the temperature is in the range of 4 to 5 $\times 10^9$ degrees and the burning yields calcium, argon, sulfur, and silicon. In the outer layers, up to $1.28M_\odot$, the temperature is still lower and the burning yields lighter elements. In the mass outside of $1.3M_\odot$, carbon and oxygen remain unburnt.

Researchers performed detailed calculations[3] for a model of one solar mass CO white dwarf which accretes mass at a rate of $4 \times 10^{-8}M_\odot$ yr^{-1}.

The results show that the explosion starts when the stellar mass reaches the value of $1.378 M_\odot$. The central density at this stage is 2.9×10^9 gm cm^{-3}. The burning front advances outward with a velocity of 0.3 of sound velocity. This event forms $0.8 M_\odot$ of elements of the iron group, of which $0.58 M_\odot$ are ^{56}Ni. A mass of $0.27 M_\odot$ consists of intermediate elements between silicon and calcium. The energy release is 1.3×10^{51} erg, most of which converts to kinetic energy.

From the theoretical calculations, we find that at least $0.5 M_\odot$ of iron forms in SN I event. This result would explain the features of the light curve of SN I. But the iron abundance in the Galaxy is lower by a factor of five than the abundance inferred from the SN I models and the estimated frequency of their occurrence. Several reasons may explain this discrepancy. It may be due to the Galaxy losing part of the iron to the intergalactic medium, or the lower frequency of SN I events than original estimates. No clear explanation has yet been found.

The agreement between calculated theoretical models and the observations of the light curve of SN I are so good that there is a tendency to use the SN I explosion as a standard candle to determine distances to far galaxies. The calculated maximum luminosity of a SN I is $5.6 \times 10^9 L_\odot$. Comparing this value to the observed luminosity of an SN I event in a distant galaxy enables us to calculate the distance to that galaxy by the decrease of luminosity due to distance.

References

1. Marschall L.A., *The Supernova Story*, 1988, Plenum Press, New York.
2. Weiler R.A., Sramek R.A., 1988, *Ann. Rev. Astron. Astrophys*, **26**, 295.
3. Woosley S.E., Weaver T.A., 1986, *Ann. Rev. Astron. Astrophys*, **24**, 205.
4. Arnett W.D., Bahcall J.N., Kirshner R.P., Woosley S.E., 1989, *Ann. Rev. Astron. Astrophys.*, **27**, 629.
5. Wilson J.R., Mayle R., Woosley S.E., Weaver T., 1986, *Ann. NY. Acad. Sci.*, **470**, 267.
6. Bethe H.A., 1993, *Astrophys. J.*, **412**, 192.
7. Barbon R., Ciatti F., Rosino L., 1979, *Astron. Astrophys.*, **72**, 287.
8. Barbon R., Ciatti F., Rosino L., 1973, *Astron. Astrophys.*, **25**, 241.

Chapter 11

Close Binary Systems

Chapter 7 presented the basic physics of a binary system. Let us now treat the specific type of binary system which is called a close binary system. In such a system each companion's presence significantly influences the other, and the evolutionary track differs significantly from that of a single star.

11.1 The Apsidal Motion

We have already mentioned the tidal effects of the companions on each other. The direct changes in the stellar structure due to the tidal effect are barely observable. However, researchers have observed the influence of the tidal effect on the synchronization of the system, resulting in changes in its orbital period. This influence can serve as a test for the theory. Such an effect is called the *apsidal motion*. Apsidal motion is actually a precession of the elliptical orbit in the binary system. Such a phenomenon is obviously easier to observe in systems which have a high degree of eccentricity. One way in which to see apsidal motion is by watching the companions in the binary system eclipse each other during their orbital motion. When the plane of rotation lies on the observer's line of sight, two eclipses can be discerned during each orbital period. If the two stars differ from each other in their luminosities or characteristic spectra, the two eclipses are distinguishable. The case in which Star 1 occults Star 2 is different from that in which Star 2 occults Star 1. Let us designate the two eclipses as the primary minimum and the secondary minimum in the common luminosity of the system. If the orbit is an ellipse, and the distances of the two stars from the centre of mass are unequal, we can expect that the time intervals between the two eclipses will not be equal. When the ellipse aligns with its major axis along our line of sight, the position of the secondary minimum will be symmetric with the two primary minima, which occur before and after the secondary minimum. If the major axis of the ellipse aligns at a right angle to our line of sight, the location of the secondary minimum will be asymmetric with respect to the two primary minima. (One can visualize this situation by drawing the elliptical orbit of the secondary star around the primary, relating the minima in the luminosity of this star to that measured by an observer located along the major axis or the minor axis of the ellipse.)

If the major axis of the ellipse rotates, there are changes in the position of the secondary minima with respect to the primary ones. The period of rotation of the major axis can thus be detected. This motion was observed in several binary systems, with period of about tens to hundreds of years.[1]

In order to formulate this problem in greater detail, we note that in a close binary system, each mass element in Star 1 is subject to the gravity of the mass of its own star and also to the gravity of the companion. The gravitational potential induced by Star 1 on its own mass elements consists of the spherically symmetric potential of the entire mass and of a perturbation potential created by the distorted distribution of the mass caused by the influence of Star 2. The potential of this star formulates as a polynomial in $\cos\theta$, where θ is the angle between the radius vector from the centre of the star to the point under consideration and the line connecting the centres of the two stars. Such a polynomial is called a *Legendre polynomial,* designated as $P(\theta)$. Due to the properties of symmetry in the system, only even powers are kept in the polynomials. The zero order term represents the spherically symmetric potential, while higher order terms represent the perturbation. The gravitational potential induced in Star 1 by the companion consists of a part which balances the centrifugal potential, and of a part which disturbs the structure of Star 1. The potential of the companion also formulates as a Legendre polynomial, where the zero order term represents the average centrifugal potential, and higher order terms represent the perturbation.

The explicit form of the perturbed potential inside a star is complicated. We are mainly interested in the mutual interaction between the companions, and the influence of the perturbation on the orbital motions. We therefore use the expression for the potential outside the stars and at their surfaces. Thus, the leading terms in the potential at a point Q at the surface

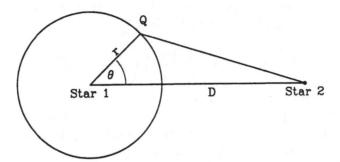

Figure 11.1. Gravitational interaction in a close binary system.

of Star 1 (see fig. 11.1) are:

$$V = -\frac{Gm_1}{r} - \frac{b}{r^3}P_2(\theta) - \frac{Gm_2}{D} - \frac{Gm_2 r^2}{D^3}P_2(\theta) \qquad (11.1)$$

where m_1, m_2 are the masses of Star 1 and Star 2 respectively, r is the distance of point Q from the centre of Star 1, and D is the separation between the companions. In this expression, the first term on the right-hand side is the regular spherically symmetric potential of Star 1. The second term is the disturbance potential created by the distortion of Star 1 caused by Star 2. The third term represents the influence of Star 2 which balances the centrifugal effects of the rotation of the system. The fourth term is the distorting potential of Star 2 which causes the tidal effects. $P(\theta)$ of the first orders are:

$$P_0(\theta) = 1$$
$$P_2(\theta) = \frac{3}{4}(\cos^2\theta - \sin^2\theta) + \frac{1}{4}. \qquad (11.2)$$

In eq. 11.1, we ignore terms with higher powers in $\cos\theta$. The term b characterizes the distortion of Star 1 from its spherical shape. It is given by:

$$b = 2KR_1^5\frac{Gm_2}{D^3} \qquad (11.3)$$

where K is a function reflecting the amount of distortion in Star 1, and R_1 is the radius of this star. The order of magnitude of K is 0.01 to 0.1.

Each mass element in Star 2 is obviously subject to exactly the same potential, with the subscripts 1 and 2 interchanged. Both stars influence the orbital motion. The dependence of the apsidal motion on the potential of the two stars can be summed up in the ratio of the period of orbital motion to the period of apsidal motion[1] (recall that the period of the orbital motion, P_{orbit}, is given by $P^2_{orbit} = \frac{4\pi^2 D^3}{G(m_1+m_2)}$):

$$\frac{P_{orbit}}{P_{apsid}} = 15 \left(\frac{K_1 R_1^5 m_2}{D^5 m_1} + \frac{K_2 R_2^5 m_1}{D^5 m_2} \right) \left(\frac{1 + \frac{3}{2}\varepsilon^2 + \frac{1}{8}\varepsilon^4}{(1 - \varepsilon^2)^5} \right) \qquad (11.4)$$

where ε is the eccentricity of the orbit.

If one of the companions in the system rotates with high angular velocity, then another term should be added to eq. 11.1 (and to eq. 11.4) to represent the centrifugal potential which acts in the star.

Recently, researchers observed apsidal motion in the system designated as "X0115 +634," which is a binary system of a neutron star and a high mass star of the type Be. Since the neutron star can be practically considered as a point mass, its distortion and influence on the apsidal motion is negligible, and apsidal motion is in this case created by the Be-star only. The system has high eccentricity ($\varepsilon = 0.34$), which enables easier detection of the apsidal motion.[2]

The ellipse rotates relative to our line of sight by 0.03^o yr^{-1}. The period of the orbit is 24.315 days. Thus we find for the ratio of P_{orbit}/P_{apsid}:

$$\frac{P_{orbit}}{P_{apsid}} = 1.802 \times 10^{-5}.$$

This value enables us to use eq. 11.3 to calculate the K of the Be-star from the observations and compare it to the theoretical calculated value. This comparison shows that the Be-star in the system must rotate with high angular velocity.

11.2 Novae

Cataclysmic variables (CVs) are short period binary systems in which a white dwarf (WD) accretes mass from a close, low mass, main sequence companion star. Most of their luminosity derives from the gravitational energy released by the accreted matter, although there are short (periodic) outbursts in which the energy comes from nuclear reactions. The WD consists of heavy elements, usually CO, while the matter accreted from the secondary is hydrogen-rich matter.

The fate of the accreted matter depends very much on the accretion rate. At a high rate of accretion, the matter heats to above the temperature threshold for nuclear reactions. It burns to helium and heavier elements in a configuration typical of a red giant star. The burnt matter is added to the original body of the WD and increases its mass gradually, up to the limiting mass for an explosion of a type I supernova (see Chapter 10, Section 10.5).

At a low rate of mass accretion, the matter does not heat up immediately and accumulates on the surface of the WD. Because of high gravity at the WD surface, the hydrogen-rich matter compresses to a very high density and becomes degenerate. A degeneracy pressure develops and forms the pressure gradient needed to support the matter against gravitation. The temperature rises slowly, and a significant amount of matter therefore accumulates before reaching the threshold temperature for nuclear reactions.

Upon obtaining this threshold temperature, the degeneracy is high, and no regulating mechanism exists between heating by nuclear reactions and pressure. Thus nuclear ignition takes place explosively. The explosion ejects the matter accumulated by accretion. A bright luminous star appears as a *nova*, or a "new star," where earlier only a very faint source could be observed (if at all) at the same location. Obviously, notwithstanding its name, a *nova* is not a "new" star, but rather a newly evidenced one. To the ancient astronomers, all "new stars" looked the same. Careful analysis of the ancient sources, however, shows that even then they noticed the difference between regular novae and the brighter supernovae.

The typical luminosity of a supernova at its maximum is about 10^{44} erg sec^{-1}, and that of a nova is about 10^{38} erg sec^{-1}. An even greater difference between them is in the total amount of released energy; in a supernova event this energy is greater by many orders of magnitude than in a nova event. This difference manifests itself both in the longer period of high luminosity, and in the greater amount of kinetic energy acquired by the ejected matter. Most of the supernova type II energy, however, is carried by neutrinos.

In a nova outburst the star ejects the outer envelope, which comprises the matter accumulated through mass accretion. The original star remains and continues to accrete mass toward the next event. In a supernova explosion the entire star explodes. A supernova type I is similar in its formation to a nova, but in the case of a supernova explosion the star vanishes entirely.

The peak luminosity reached in the nova outburst is the Eddington luminosity, which is about 20,000 solar luminosities. The envelope of the star is ejected at a velocity of few hundred to a few thousand km sec^{-1}. The amount of ejected matter is about the same amount as the accreted matter, which is around $10^{-5} M_\odot$. This ejected matter can be observed for several years as an expanding cloud around the star, until it fades into the interstellar medium.

Astronomers categorize classical novae by the rate of decline in their luminosity. A quantity t_2 is the time required for nova luminosity to decrease by a factor of approximately six (two magnitudes) below its maximum. For very fast novae t_2 is a few days, for fast novae t_2 is 10 to 50 days, and for slow novae 80 to 250 days.

A special class consists of the so-called *dwarf novae*, which are entirely different objects. Such objects burst frequently with periods of days to yr. During an outburst, their luminosity reaches values of 10 to 100 times their quiescent luminosity. The energy source of their outburst is not nuclear reactions but a sudden release of gravitational energy, caused by a very high rate of mass accretion.

The overall picture of a nova explosion is of a recurrent event in which, during a quiescent phase, hydrogen-rich matter accretes onto a white dwarf. During the short period of an outburst, the star ejects this matter explosively. Prialnik[3] carried out an evolutionary calculation of a complete cycle. In this evolutionary sequence, the initial mass of the WD is $1.25 M_\odot$. The WD consists of carbon and oxygen (CO), covered by a very thin hydrogen-rich envelope of $5 \times 10^{-8} M_\odot$. The temperature at the base of this envelope is 5×10^6 degrees, and the density is 100 gm cm^{-3}.

During a quiescent phase, which lasts for 5×10^5 yr, the WD accretes mass at a rate of $10^{-11} M_\odot$ yr^{-1}. The accreted matter does not heat up sufficiently to ignite nuclear reactions. It becomes degenerate and heats up very slowly. During the long quiescent phase, a small amount of hydrogen from the accreted matter diffuses into the outer layers of the CO core, and carbon and oxygen diffuse into the envelope. When the temperature at the bottom of the hydrogen-rich layer reaches the threshold for nuclear reactions, these reactions ignite explosively, creating a very high luminosity of up to $10^6 L_\odot$. Such high luminosity cannot be transferred by radiation. Convection develops which carries the high energy flux and mixes the material from the bottom of the burning hydrogen shell upward. The temperature rises to 2×10^8 degrees, and lifts the degeneracy. With the rising temperature, the envelope begins to expand, and a shock wave forms that ejects part of the envelope matter. The amount of mass ejected by the shock wave is about $1.2 \times 10^{-6} M_\odot$, which is one-sixth of the envelope mass. It is ejected at a velocity of 3,800 km sec^{-1}. This stage lasts for about six days.

The radius of the star increases at this stage to $100 R_\odot$, with an effective temperature of 9,000 degrees. The luminosity is the Eddington luminosity, and further mass loss from the star occurs in the form of a thick wind driven by the radiation pressure. The velocity of the wind is about 100 km sec^{-1}. This stage lasts for 17 days, and the amount of mass ejected during this stage is $3.8 \times 10^{-6} M_\odot$.

At the end of this stage, the envelope becomes transparent. The star contracts, and, on contracting, a mass shell of $1.3 \times 10^{-6} M_\odot$ remains bloated. This shell has a velocity higher than the escape velocity. It leaves the star with a velocity of 30 to 50 km sec^{-1}. The total amount of mass ejected in the three stages is $6.3 \times 10^{-6} M_\odot$, of which $5 \times 10^{-6} M_\odot$ is the accreted mass and the rest of the mass comes from the original body of the WD. The high luminosity of the star lasts for 25 days, a period characteristic of a fast nova. During this period the WD itself has warmed up. After the mass ejection, the star reverts to nearly its original state, but is a little hotter and has a slightly larger radius than it had initially. During a few more hundreds of years the star cools down to its original state, and the cycle of mass accretion begins again. It is believed that such a cycle may repeat itself some 10^4 times. Through each event, the mass of the WD decreases by a $10^{-6} M_\odot$. During the quiescent phase between subsequent events, the star cools down by means of the routine cooling process of a WD.

This evolutionary sequence is evidently an example whose numerical results depend upon the initial parameters of the model. The same calculations can be used for a wide range of parameters that will result in the great variety of nova outbursts found in nature.

We find that the crucial parameters determining the event of a nova outburst are the mass of the WD and the rate of the mass accretion. If the mass of the WD is low, the gravity at its surface is low as well, and high degeneracy does not develop in the accreted matter. In the absence of high degeneracy, no explosive ignition will start. The lower limit for the WD mass that can ignite a nova outburst is $0.8 M_\odot$.

The rate of mass accretion is important because at a high rate of mass accretion the accrued matter heats up immediately and starts nuclear burning in a nondegenerate state. The burnt matter is added to the original body of the WD. The upper limit on the rate of mass accretion for nova outbursts is between 10^{-9} to $10^{-8} M_\odot$ yr^{-1}. The lower the rate of the accretion, the stronger is the outburst that follows, at the cost of requiring a longer quiescent period for the accumulation of matter. A low rate of mass accretion demands longer accretion time because the WD mass determines the amount of the accumulated mass needed to trigger the outburst. However a longer quiescent period allows for a deeper diffusion resulting in a stronger outburst, as is explained in the next paragraph. Accretion rates lower than

$10^{-12} M_\odot$ yr^{-1} demand very long quiescent periods. These circumstances do not accord well with the statistics of nova outbursts in the Galaxy.

One of the important features in the creation of a nova outburst is the process of diffusion taking place during the quiescent phase. In theoretical calculations we find that when the burning material is hydrogen-rich matter containing only few percent of metals, the burning process is relatively slow and cannot explain the evolution of a fast nova such as the one described above. In order to maintain a sufficiently high rate of nuclear energy production for a fast nova outburst, the system requires a high percentage of CNO elements.

Most observations of nova outbursts show a high abundance of heavy elements. Researchers began to search for the source of these heavy elements, which were essential in both theory and observation. Clearly the low mass, main sequence star which donates the accreted mass cannot be the source of these elements.

They found[4] that the diffusion of elements between the hydrogen-rich envelope and the CO body of the WD during the quiescent period is responsible for the presence of heavy elements in the burning matter. A very steep gradient in the composition exists at the boundary surface between the two regions. This gradient drives the diffusion. The process of diffusion is very slow, and the amount of transferred matter is very small, especially of the heavy elements. The hydrogen diffused into the CO region reaches a value of few percent at a depth (in mass) of $1.2 \times 10^{-6} M_\odot$ below the boundary surface.

However when the nuclear reactions begin, this diffused hydrogen is the first to ignite. Because of the high temperature and the steep temperature gradient that forms, an efficient convection develops which mixes the matter above this point, up to the outer surface of the envelope. Thus $1.2 \times 10^{-6} M_\odot$ of almost pure CO matter mixes into the $5 \times 10^{-6} M_\odot$ of the envelope, and raises the fraction of heavy elements to 0.25 of the matter. This enrichment appears also in the ejected matter. A low rate of mass accretion demands longer periods of time between consecutive outbursts. This low rate allows deeper and more effective diffusion of hydrogen into the CO body, thereby producing stronger outbursts.

We mentioned earlier that we expected the peak luminosity of a star to be its Eddington luminosity. However observations of fast novae find that for short periods at the beginning of the outbursts, luminosity peaks reach super-Eddington levels of up to 10 times or more the Eddington limit.

We recall that the definition of an Eddington luminosity assumes the existence of a hydrostatic equilibrium and radiative transfer of the energy in a spherically symmetric configuration. Undoubtedly during the first stage of the nova explosion, no hydrostatic equilibrium exists in the exploding envelope. Therefore, the evolution takes place on a dynamical time scale, and the hydrostatic equation does not hold. The strong and efficient convection formed by the steep temperature gradient carries most of the energy released at the envelope's bottom and mixes matter from the bottom layers up to the stellar surface.

We already mentioned that convection results in the presence of a high abundance of heavy elements in the envelope of a fast nova. When the degeneracy is lifted, the temperature is above 10^8 degrees, whereby the dominant burning process becomes the CNO cycle. At such a high temperature, the characteristic time for the proton capture (see tab. 3.2) shortens to a few seconds. The CNO cycle includes a positron decay of ^{15}O. Since temperature does not influence the rate of positron decay, this process becomes the bottleneck of the entire cycle. The characteristic time for this decay is 176 seconds. However the convection is so vigorous that these unstable elements reach the surface before the decay takes place. Thus the decay stage occurs at the surface, and the energy released in this stage serves to accelerate the outer layers of the star, which the star eventually ejects.

In evolutionary calculations of a fast nova,[5] the efficiency of the convection used in the computer evolutionary code influences the luminosity significantly. We recall from eq. 2.26 the free parameter α, which is the ratio of the mixing length to the pressure scale height. In the calculations referred to here, we find that increasing α from 0.1 to 1 enhances the peak luminosity by a factor of 20. Evidently, efficient convection influence both the energy transfer and the transfer of the unstable elements to the surface, where they release their nuclear energy.

The mechanism of the nova outburst is very well understood. It is commonly agreed that the event takes place in a close binary system. However the existence of the binarity is well established only in a very few novae. As an example of a concrete investigation of a nova system, let us consider the following event.

On 23 March 1991 a nova designated as Nova Herculis 1991 appeared in the constellation Hercules. The progenitor star was known from earlier observations, and the peak luminosity at outburst was more than 10^5 times the original stellar luminosity. A team of astronomers[6] from the Wise Observatory at Tel-Aviv University followed this star, after its eruption, with the aim of revealing details of the system. During a period of twelve weeks

they observed the star frequently, and the results of these observations are
of considerable interest. During this period the luminosity faded by a factor
of 16 and became bluer, which means that the radius of the primary star
decreased and its effective temperature rose. They observed a periodical
dip in the luminosity of the system, with a period of 0.29764 days (7.1433
hours). The width of the dip shoulders is about three hours. This width
remained constant throughout the whole observational period, while the
depth increased. The bottom of the dip is not flat. From the relation between the luminosity and the effective temperature, they found that the
radius of the primary star decreases by a factor of about five. The observations had to be interrupted in October 1991 because the star became a
"day's object," and could not be observed for the next half year. Figure 11.2
displays the light curve of Nove Herculis 1991.

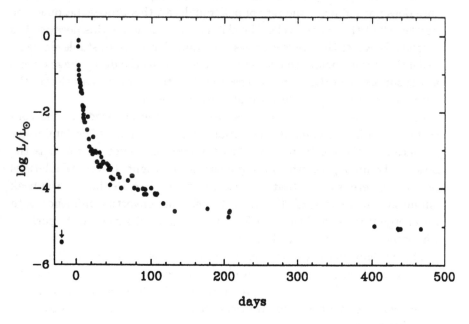

Figure 11.2. Light curve of Nova Heculis 1991. The point marked by
an arrow (at day −20) designate stellar luminosity before the outburst as estimated from "Palomar Sky Survey Atlas," compiled a
number of years ago. The gap between day 207 and day 403 corresponds to the period in which the star could not be observed beacuase
it became a "day's object." [Professor E. Leibowitz kindly supplied
this data, where a part was collected by the team at Wise Observatory, and the other part was extracted from IAU Circulars.]

From these results, the team concluded that some component in the system is eclipsed, and they inferred the following details: The fact that the minimum in the luminosity is not flat means that the eclipse is partial. If the primary star is the eclipsed star, then the considerable change of its radius would have caused the eclipse either to vanish or to become total. For the same reasons, the secondary star cannot be eclipsed by the primary companion. Thus they conclude that the dip in the luminosity is due to the eclipse of the accretion disk in the system by the secondary star, whose radius does not change during the nova event.

The depth of the dip increased during the observational period because, due to the decrease in the luminosity of the primary star, the weight of the luminosity of the disk increased in the total luminosity of the system. The structure of the disk and the orbital period depend on the dynamical parameters of the system, and these remain constant even when the primary star undergoes significant changes in its features. Hence the width of the dip in the luminosity remains constant, while the total luminosity changes significantly. At the beginning of the observations, immediately following the nova outburst, the dip was hardly noticeable. This may have been the case either because the disk was destroyed during the eruption and was later recovered, or because the depth of the dip was too shallow to be discerned at the start of the observation period. The choice between these two possibilities can be made only after obtaining more details about the system. Researchers believe that collecting more observational data about the system, such as the relative velocities of the companions as determined by the Doppler shift of their spectra, may eventually reveal the complete physical structure of the system.

11.3 Mass Transfer in a Close Binary System

When a star in a binary system loses mass by wind or mass ejection, part of the lost mass will accrete to the companion. In close binary systems, however, an active mass transfer can occur in which the rate of mass transfer depends on the presence of the close companion. When one star loses mass because it overflows the Roche-lobe, most of the lost mass accretes directly to the companion. The character of the mass transfer and the response of the stars to this process depend very much on the distance between the companions, on the mass ratio of the companions, and on the evolutionary phase of each one of the stars.

In a close binary system where the primary star is a compact object such as a neutron star, we distinguish between the case in which the mass of the secondary star is lower than the primary's mass (called a *low mass binary* or LMB) and that in which the mass of the secondary is higher than that of the primary (called a *high mass binary* or HMB).

The formation of an LMB system appears to be straightforward. The higher mass star in the system evolves faster than its companion. If its initial mass is sufficiently high it may pass through the entire evolutionary track, up to the explosion of a type II supernova which creates a neutron star. Meanwhile, the lower mass star remains at the beginning of its main sequence phase. In such a case we must assume that all the dramatic transitions through which the primary went had little influence on the evolutionary track of the secondary.

The evolution of the HMB system demands a more complex evolutionary track. The primary must start as the higher mass star in order to complete its evolutionary track a great deal ahead of the secondary. However during its evolution the primary has lost a significant amount of mass, part of which accretes to the secondary, increasing the secondary's mass significantly above its initial value. Thus when the primary completes its evolutionary track to become a neutron star, the secondary, having evolved more slowly due to its lower initial mass, now has a higher mass than the primary, although its evolutionary phase is delayed.

A more complicated evolutionary track exists for close binary systems in which the initial masses of the companions did not differ significantly. In such systems, the companions exchange the roles of mass donor and mass accretor. Consider a situation, for example, in which the heavier of the companions evolves faster. When it becomes a red giant and expands, the heavier star overflows its Roche-lobe and sheds mass to its companion. If the amount of transferred mass is significant, the lower mass star may become much more massive than the initially higher mass star. Its evolution therefore accelerates to a point where it may overtake the evolution of the initially advanced star. Such scenarios may repeat several times, with the roles of the companions changing in the process. We go no further in setting up such scenarios, but we need only use our imaginations to consider all the possible alternatives.

Close binary systems in which one of the components is a neutron star (NS) are sources of X-ray radiation. The radius of a neutron star is about 10 kilometres, which is few times its Schwarzschild radius. A particle accreted to such an object will, prior to its accretion, radiate the gravitational energy released while falling to the potential well of the neutron star. As it turns out, the wavelength of the radiation in such a case is in the range of X-rays.

Most X-ray radiation cannot pass through the Earth's atmosphere. Until recently the observations within this range of wavelengths were confined to those carried out by balloon-mounted and rocket-mounted telescopes. Owing to the limited facilities of such devices, researchers obtained only a vague picture of the distribution of X-ray radiation over the sky. Except for a small number of prominent point sources as SCO X-1, Cyg X-1, and a few others, the accepted picture was of a scattered and diffused distribution of such radiation.

During the last two decades satellite-mounted telescopes have significantly widened observational facilities within this range of wavelengths, finding that most X-ray radiation originates in point sources. To give an idea of the rapid advances being made in this field, we need only note that a catalog published in 1983[7] includes 115 identified X-ray sources in our Galaxy and in the nearby Magellanic Clouds; 65 of these 115 sources are believed to be close binary systems which include a neutron star as one of the components. A catalog being prepared now,[8] however, categorizes 113 sources as close binary systems with a neutron star as one of the components.

An important factor in the evolution of close binary systems is the loss of angular momentum by *gravitational radiation* (GR) and *magnetic braking* (MB). A rotating binary system forms a gravitational quadrupole which radiates gravitational radiation. This radiation leaves the system with a certain of angular momentum, and through it angular momentum is lost from the system. The loss of angular momentum by gravitational radiation in c.g.s. units is given by:

$$\left(\frac{\dot{J}}{J}\right)_{GR} = -\frac{32}{5}\frac{m_1 m_2}{c^5}\frac{G^3 M}{D^4} \tag{11.5}$$

where m_1, m_2 are the masses of the primary and secondary stars respectively, M is the total mass of the system, and D is the separation. Note the high (inverse) dependence of this expression on the separation (D^4). This process thus becomes important only in very close systems.

The magnetic fields existing in a star also cause loss of angular momentum by the effect known as *magnetic braking*. The reasons for this loss of angular momentum are similar to those considered for magnetic braking of a single star (see Chapter 9).

The expression for the angular momentum loss due to magnetic braking, MB, is:[9]

$$\left(\frac{\dot{J}}{J}\right)_{MB} = -5 \times 10^{-30}\frac{R_2^4 G M^2}{f^2 m_1 D^5}. \tag{11.6}$$

The symbols have the same meaning here as in eq. 11.5; R_2 is the radius of the secondary star and f is a function on the order of unity whose dimensions are of length divided by the square root of time: $[f] = [R/t^{1/2}]$.

Since a close binary system loses angular momentum according to the eqs. 11.5 and 11.6, the orbital separation between the companions decreases, and such systems evolve toward smaller separation. When the separation in the system decreases, the Roche-lobes of the companions decrease as well. The radius of the neutron star (NS) companion is very small relative to its Roche-lobe radius; but the radius of the secondary, which is a main sequence star, is of the order of magnitude of its Roche-lobe radius. When the Roche-lobe radius decreases, the star may overflow its Roche-lobe, and begin to transfer mass to the NS. For systems in which the secondary's mass is lower than that of the primary, the ratio of the Roche-lobe radius, R_L, of the secondary to the orbital separation, D, is given by:

$$\frac{R_L}{D} = \frac{2}{3^{4/3}} \left(\frac{m_2}{M}\right)^{1/3}. \tag{11.7}$$

The transfer of mass between the companions results in a new distribution of angular momentum in the system, and this new distribution may lead to changes in the separation, D. If in the course of the mass transfer a part of the mass leaves the system, this mass may carry angular momentum and thus further decrease the angular momentum of the system. In the overall picture, the changes in the separation are due to the loss of angular momentum by GR and MB, and to the redistribution of the angular momentum in the system inferred from the mass transfer.[10] This process is summed up in the expression for $\dot{D}$ (the rate of change for the separation):

$$\frac{\dot{D}}{D} = 2\left(\frac{\dot{J}}{J}\right)_{GR} + 2\left(\frac{\dot{J}}{J}\right)_{MB}$$

$$- \frac{\dot{m}_2}{m_2}\left[2\left(1 - \frac{\beta}{q}\right) - \left(\frac{1-\beta}{1+q}\right) - \frac{2\alpha(1-\beta)(1+q)}{q}\right] \tag{11.8}$$

where β is the fraction of the mass ejected from the secondary and accreted to the primary, $1-\beta$ is the mass fraction lost from the system, q is the ratio of the primary's mass to the mass of the secondary ($q = m_1/m_2$), and α is the specific angular momentum carried away by the mass lost from the system, in units of $2\pi D^2/P_{orbit}$.

Note that $\dot{J}_{MB}$, $\dot{J}_{GR}$, and $\dot{m}_2$ are negative. This means that the first two terms on the right-hand side of eq. 11.8 represent a decrease in the separation, while for systems in which $q > 1$, ($m_1 > m_2$), the third term represents an increase in the separation. Thus when the decrease in the separation due to MB and GR results in mass transfer (due to decrease in the Roche-lobe radius and overflow of the secondary), the third term represents an increase in the separation caused by this mass transfer. A negative feedback mechanism is established between the mass transfer and the decrease of the separation which drives this mass transfer. This feedback mechanism controls the overall process to approach a rate at which the increase of the orbital separation due to mass transfer will not overcome the separation decrease due to GR and MB.

11.4 LMXB System

An LMXB system is a close binary system in which the primary is a neutron star with a mass of around $1.4M_\odot$, and the companion is a low mass star in its main sequence evolutionary phase.

When the secondary overflows its Roche-lobe, the neutron star (NS) attracts the mass in excess of the secondary's Roche-lobe radius. Ordinarily the transferred matter has a certain amount of angular momentum, and does not accrete in a head-on collision. Instead it forms a disk (an accretion disk) around the NS. Because of the viscosity in the disk, the kinetic energy and angular momentum of the matter particles decrease, and the matter is gradually accreted to the NS. The accretion, produces high energy radiation which is usually in the form of X-rays, and such systems furnish one of the observed X-ray sources.

Researchers carried out[11] evolutionary calculations of low mass X-ray binary systems with the aim of following the evolution of the system's parameters according to eqs. 11.5, 11.6, and 11.8. The system we shall now consider contains a compact star (NS) and a low mass main sequence star with a mass of $0.4M_\odot$. Due to loss of angular momentum by GR and MB, the orbital separation of the system decreases gradually while a decrease in the Roche-lobe of the secondary follows. This occurs until mass transfer from the secondary to the NS commences due to the overflow of the Roche-lobe by the secondary. The process of mass transfer and decrease in orbital separation continues until very short periods of 70 minutes, with the secondary mass being about $0.08M_\odot$.

At the beginning of the evolution, the time scale for the mass decrease of the secondary is on the order of 10^9 yr, which is longer than the thermal time scale of the star. Following the mass loss, the star has enough time to relax to a new thermodynamic equilibrium with lower entropy. The decrease in the stellar mass results in a decrease in the stellar radius.

When the orbital separation becomes very small, the rate of the GR angular momentum loss increases significantly (note the $1/D^4$ dependence of expression 11.5), and the time scale for the mass transfer becomes much shorter than the thermal time scale. In this situation, the star does not have enough time to relax to a new thermal equilibrium, and its evolution is almost adiabatic, with no change in the specific entropy. Recall eq. 4.7, where we found that through adiabatic changes of a polytropic configuration, the change of the radius is in a direction opposite to the change of the mass. Upon reaching this point, further evolution includes a decrease in stellar mass due to Roche-lobe overflow, followed by an increase in the radius. The regulating mechanism represented by eq. 11.8 causes an increase in the separation and the orbital period. There is thus a minimal period in close X-ray binaries.

Variations in the stellar parameters or the physics used in calculations, like the variations in stellar composition or the specific equations calculating the opacity and the equation of state, result in variations in the exact value for the minimal period. However it is clear that there is a minimal period of 65 to 80 minutes for an evolutionary sequence such as the one described above. This result fits in nicely with the observational finding that low mass X-ray binaries hardly ever have orbital periods shorter than one hour. (There are two exceptional cases, one with an orbital period of 11.4 minutes and the other with an orbital period of 50 minutes. We must consider a specific evolutionary conditions for their formation.)

The evolution of the secondary continues until it relaxes to a new thermal equilibrium with a smaller radius. Meanwhile the orbital separation and the secondary's Roche-lobe increase, and the contracting star loses contact with its Roche-lobe and mass transfer ceases. The specific results of further evolution depend on the detailed properties of the system at this stage.

Part of the X-ray radiation hits the secondary and heats it up. As a result of this heating, the secondary expands. Since it is already on the limit of its Roche-lobe, it will overflow the Roche-lobe further and shed more mass to the primary. This mass transfer creates further X-ray radiation which will further heat the secondary. Thus a positive feedback forms between the mass transfer and the heating. The X-ray radiation resulting from this mass transfer causes the heating of the secondary. Three mechanisms modify this feedback however: (1) the Eddington luminosity of the NS; (2) shadowing effect by the accretion disk; (3) an increase of the separation due to the mass transfer.

The Eddington Luminosity of the NS.

When X-ray radiation produced by the mass transfer exceeds the Eddington luminosity of the NS, the radiation pressure rejects the mass transferred by the secondary from the system, and this mass does not accrete to the primary. The rate of mass accretion which produces the Eddington luminosity therefore limits the rate of mass accretion onto the NS . Recalling eq. 3.15, Eddington luminosity, L_{Edd}, is given by:

$$L_{Edd} = \frac{4\pi cGM}{\kappa}. \tag{11.9}$$

Inserting for M the mass of the NS $\simeq 1.4M_{\odot}$, and $\kappa \simeq 1$ cm^2 gm^{-1}, we find that for the NS, $L_{Edd} = 7 \times 10^{37}$ erg sec^{-1}, which is about 20,000 solar luminosities. The amount of energy released by a mass unit falling to the surface of the NS equals the potential difference between the stellar surface and the initial location of the mass. The absolute value of the potential at the initial location of the mass is very low relative to the absolute value of the potential at the NS surface. Hence the potential difference equals approximately the potential at the NS surface, V_{NS}:

$$V_{NS} = -\frac{GM}{R}.$$

The rate of energy release by the mass accreted at a rate of $\dot{m}$ is:

$$\frac{dE}{dt} = -\dot{m} \cdot V_{NS} = \frac{GM\dot{m}}{R} \tag{11.10}$$

where M is the NS mass, R is its radius, and $\dot{m}$ is the rate of mass accretion. We can compare this expression to eq. 11.9 to find that the rate of mass transfer which creates the Eddington luminosity, is $\dot{m}_{Edd} \simeq 5 \times 10^{17}$ gm sec$^{-1} \simeq 10^{-8} M_{\odot}$ yr^{-1}. This is an upper limit for the rate of mass accretion onto the NS. Any mass leaving the secondary in excess of this limit will leave the system into the interstellar medium.

Shadowing by the Accretion Disk

The accretion disk formed around the NS resides in the plane of the orbital rotation of the system. The secondary star is also located in the same plane. Thus the accretion disk shadows the secondary from the X-ray radiation, which is created very close to the NS surface. The degree of shadowing depends on the geometry of the system (namely the ratio R_2/D) and on the thickness of the accretion disk, which increases with the rate of the mass transfer. A high rate of mass transfer thickens the disk and strengthens its shadowing, thereby lowering the amount of heating luminosity which reaches the secondary. When the mass transfer decelerates, the disk becomes thinner and a larger fraction of the heating radiation can reach the secondary. The shadowing effect of the accretion disk forms a negative feedback which stabilizes the heating process and the mass transfer rate around some medial value. Unfortunately there is as yet no established theory for the factors which govern the structure of thick accretion disks. Only first approximations are used to calculate the mutual influence between the accretion disk and the heating of the secondary.

The Increase of the Separation due to Mass Transfer

When mass is transferred from the secondary to the primary, it also carries angular momentum from the former to the latter. In LMXB, the distance of the primary from the centre of mass around which the masses rotate is smaller than that of the secondary. Each mass element moving from the secondary to the primary has an excess of specific angular momentum relative to the specific angular momentum of the mass elements located there. There is a redistribution of this excess angular momentum in the system, which causes the separation between the companions to increase. Thus an increase in the orbital separation of the system follows the mass transfer. This increase is calculated by the last term of eq. 11.8. The enlargement in separation implies an increase in the Roche-lobe radius of the secondary, according to eq. 11.7. This increase moderates the overflow of the secondary beyond its Roche-lobe due to the heating.

Detailed evolutionary calculations of close binary systems according to eqs. 11.5, 11.6, and 11.8 obtained a clear picture of the mutual interaction between mass transfer due to Roche-lobe overflow and heating by the X-ray radiation created by this mass transfer. The calculations included the modifications of this process mentioned above.

The response of the secondary star to the heating depends very much on its initial state before the heating.[12] The response of higher mass stars, with masses of one solar mass or more, is moderate and their radius increase is limited. On the other hand the response of lower mass stars is very strong. The increase in their radius is very large and takes place on a short time scale.

The cause for this different behaviour is the width of the neutral zone in the star before heating. When a star is heated by an external source, it responds by creating a new structure capable of re-radiating the heating luminosity plus the energy produced in its interior. The radiation from the stellar surface depends on its radius and on the effective temperature: $L = 4\pi R^2 \sigma T_e^4$, where T_e and R are the effective temperature and the radius of the star, respectively. The amount of heating luminosity absorbed equals the cross-section of the star, πR^2, times the flux of the heating luminosity, F_{heat}. Comparing the two expressions, we find that for re-radiation of the heating luminosity to occur, the effective temperature must satisfy the relation:

$$T_e^4 = \frac{F_{heat}}{4\sigma}. \qquad (11.11)$$

In obtaining this relation, we ignored the intrinsic stellar luminosity under the assumption that it is negligible relative to the heating luminosity. The characteristic heating flux by X-ray radiation in a close LMXB is around 10^{12} erg sec^{-1} cm^{-2}. Re-radiation of the luminosity that heats the star at this flux demands an effective temperature of above 10,000 degrees. At this effective temperature, the star fully ionizes. This means that any neutral zone existing in the star before the heating should now be ionized.

When a sample of matter ionizes, it goes through two changes which are important for the stellar structure. First, with ionization the pressure increases due to the increase of the number of particles per unit mass, $1/\mu$ (recall eq. 3.7). Second, the opacity of neutral matter is very low, and with ionization the opacity increases by few orders of magnitude. Both these changes act to increase the radius: the first change is responsible for an instantaneous expansion, while the increase of opacity has a longer time scale.

Higher mass stars, which are hotter, have narrow neutral zones and respond with moderate expansion to the heating. However, low mass stars, which initially have low effective temperatures and wide neutral zones, respond with a greater and more violent expansion.

In fig. 11.3 we display the evolutionary track of a low mass secondary $(M = 0.4 M_\odot)$ when it starts overflowing its Roche-lobe in a close LMXB system. It is heated by the X-ray radiation created by the mass transfer. The lower panel displays the stellar mass which decreases along this evolutionary track. The second panel from the bottom displays the expansion of the stellar radius due to the heating, which causes the further overflow of the Roche-lobe. The third panel displays the increase of the orbital period, which reflects the increase in separation between the companions due to the mass transfer. The fourth panel displays the luminosity of the star, which increases gradually until it levels with the heating luminosity. The top panel displays the rate at which the mass transfers from the secondary. It is evident that the rate of mass transfer is higher than $\dot{m}_{Edd}$ for the

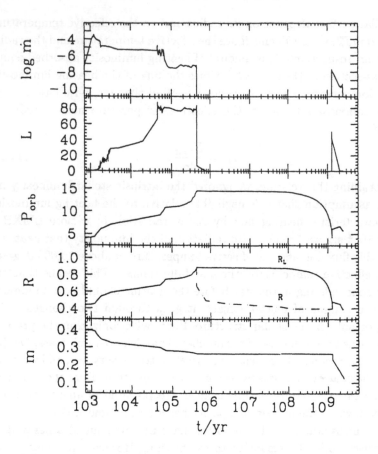

Figure 11.3. Evolution of an LMXB with heating X-radiation. [Adopted from Harpaz and Rappaport.[12]]

first 4×10^4 yr. Only when the star reaches the effective temperature that enables it to re-radiate the heating luminosity does the evolutionary track becomes more moderate.

The star cannot achieve the higher values of effective temperature earlier because, while its surface layer heats, that same mass layer is simultaneously being removed from the star by the Roche-lobe overflow. From the figure we learn that mass transfer takes place in the system for about 4.5×10^5 yr. At the end of this period, the secondary star loses contact with its Roche-lobe, and mass transfer ceases. Due to the cessation of X-ray heating, the stellar radius shrinks (dot-dashed line in the second panel from the bottom). Only after another $\sim 10^9$ yr do the separation and the Roche-lobe decrease to such values that enable renewal of mass transfer.

References

1. Schwarzschild M., 1958, *Structure and Evolution of Stars*, Princeton University Press, Princeton.

2. Tamura K., Tsunemi H., Kitamoto S., Hayashida K., Nagase F., 1992, *Astrophys. J.*, **389**, 676.

3. Prialnik D., 1986, *Astrophys. J.*, **310**, 222.

4. Kovetz A., Prialnik D., 1985, *Astrophys. J.*, **291**, 812.

5. Hayes J., Truran J.W., Livio M., Shankar A., 1990, in *Accretion-Powered Compact Binaries*, ed. C.W. Mauche, Cambridge University Press, Cambridge.

6. Leibowitz E.M., Mendelson H., Marshal E., Prialnik D., Seitter W.C., 1992, *Astrophys. J. Lett.*, **385**, L49.

7. Bradt H.V.D., McClintock J.E., 1983, *Ann. Rev. Astron. Astrophys.*, **21**, 13.

8. Remillard R.A., Bradt H.V.D., Schwartz D.A., Tuohy I.R., 1993, to appear in *Astrophys. J. Supp.*

9. Verbunt F., Zwaan C., 1981, *Astron. Astrophys.*, **100**, L7.

10. Rappaport S.A., Joss P.C., Webbink R.F., 1982, *Astrophys. J.*, **254**, 616.

11. Rappaport S.A., Verbunt F., Joss P.C., 1983, *Astrophys. J.*, **275**, 713.

12. Harpaz A., Rappaport S.A., 1991, *Astrophys. J.*, **383**, 739.

Chapter 12

Special Topics

12.1 Solar Neutrinos

Most nuclear reactions which produce nuclear energy in stellar interiors involve β decay. The usual process of β decay is one in which the unstable neutron decays to a proton while emitting an electron and an antineutrino. Under special conditions the inverse process takes place. A proton converts into a neutron while getting rid of the extra electric charge by emitting a positron and keeping the spin balance of the system by emitting a neutrino. The cross-section for the interaction of a neutrino with baryonic matter is negligible, and, for practical purposes, an ordinary star is transparent to neutrinos.

Most of the energy produced in the nuclear reactions is in the form of photons, which are immediately absorbed in the stellar matter. Their energy adds to the thermal reservoir of the star. This energy is responsible for the formation of the pressure gradient needed to support the star against gravity and for the radiation of stellar luminosity. The energy carried by the neutrinos radiates into the interstellar medium directly from the point where it formed; it is not involved in the energy balance of the star and is not included in the observed stellar luminosity. Ordinarily, when we calculate the energy balance in stars, we subtract the energy carried by the neutrinos directly from the amount of energy produced by nuclear reactions.

Because the neutrinos radiate directly from the point at which they formed, they carry information about this location. Observation of the neutrinos may thus reveal direct information about the stellar interior. The hope is that by observing neutrinos from the Sun, which is the closest

211

star to our own location, we may be able to compare theories of stellar structure with a concrete example of a solarlike star.

Due to the small cross-section for interactions of neutrinos with matter, it is very difficult to observe these particles. We need special detectors therefore to detect neutrinos and measure their energies.

The first neutrino detector proposed for astrophysical use employed chlorine as the detecting material. The idea was that when a chlorine atom absorbs a neutrino, the atom converts into an argon atom by emitting an electron:

$$^{37}\text{Cl} + \nu \longrightarrow \ ^{37}\text{Ar} + e^-. \tag{12.1}$$

This conversion is an endoergic reaction. The transition from a ground state of ^{37}Cl to the ground state of ^{37}Ar demands absorption of energy of 0.814 MeV. A more favourable transition is from the ground state of ^{37}Cl to an excited state of ^{37}Ar, where the first excited state of ^{37}Ar is 1.41 MeV above its ground state. This transition therefore demands neutrinos with minimal energy of 0.814 MeV and favours neutrinos with energies of 2.224 MeV and above.[1] Table 12.1 displays the processes of hydrogen burning that release neutrinos, with the involved neutrino energies and the fluxes of these neutrinos predicted at the Earth's surface. The fluxes are calculated for the conditions supposed to exist at the solar centre, with a temperature of 15.6 million degrees and a density of 148 gm cm^{-3}. This model is called the standard model of the Sun and assumes the following: the Sun is in thermal and hydrostatic equilibrium; there are no deviations from spherical symmetry; in the inner part of the Sun (up to $0.985 M_\odot$ in mass and $0.72 R_\odot$ in radius) energy transfer occurs by radiation; and the composition observed at the solar surface is the initial composition of the Sun.

Note that the p-p reaction has a variation in which an electron is absorbed instead of a positron emitted. This variation is called p-e-p (proton, electron, proton) reaction. The neutrino emitted in this variation is more energetic, but this variation is much less favourable in the conditions existing at the solar centre.

Comparing the neutrino energies displayed in tab. 12.1 with the energies needed to activate the chlorine conversion to argon, we observe that the energy of the neutrinos released in the main p-p reaction, PPI, is below the minimal energy that the chlorine detector can measure. The neutrinos released in the other channels have energies of around or above the minimal energy spotted by the chlorine detector. Only one of them, PPIII, has

Table 12.1. Neutrino energies and fluxes (energies given in MeV).

process				Max. neut. energy	av. neut. energy	neut. flux cm^{-2} sec^{-1}
PPI	$^1H + {}^1H$	$\Rightarrow$	$^2D + e^+ + \nu$	0.420	0.263	6.0×10^{10}
PeP	$^1H + e^- + {}^1H$	$\Rightarrow$	$^2D + \nu$	1.442		1.4×10^8
PPII	$^7Be + e^-$	$\Rightarrow$	$^7Li + \nu$	0.862	0.80	4.7×10^9
PPIII	^8B	$\Rightarrow$	$^8Be + e^+ + \nu$	14.02	7.2	5.8×10^6
CNO	cycle					
	^{13}N	$\Rightarrow$	$^{13}C + e^+ + \nu$		0.71	6.1×10^8
	^{15}O	$\Rightarrow$	$^{15}N + e^+ + \nu$		1.04	5.2×10^8
	^{17}F	$\Rightarrow$	$^{17}N + e^+ + \nu$		0.94	5.2×10^6

energies that are well above the threshold allowing for the favourable channel of conversion of chlorine to argon.

The last column in tab. 12.1 presents the expected fluxes of the neutrinos produced in the different channels of the standard model in units of neutrino number cm^{-2} sec^{-1} at the Earth surface. The cross-sections for neutrino interaction with chlorine atoms depend on their energies; tab. 12.2 gives the values of the cross-sections, averaged over the experimental spectra, for the neutrinos released in the different channels. (Also included in the table are the cross-sections for the interaction of neutrinos with gallium and hydrogen, which are used in detectors discussed later.)

Table 12.2. Cross-sections for neutrinos, in units of 10^{-46} cm^2.

	PPI	PeP	^7Be	^8B	^{13}N	^{15}O	^{17}F
^{37}Cl	0.	16	2.4	1.06×10^4	1.7	6.8	6.9
^{71}Ga	11.8	215	73.2	2.43×10^4	61.8	116	117
^2H	0.	0.	0.	1.03×10^4	0.	0.	0.

Multiplying the cross-sections given in tab. 12.2 by the fluxes given in tab. 12.1, we find that in order to obtain a reasonable value of interaction for neutrinos with chlorine (like 10^{-5} sec^{-1}), we must have an amount of about 10^{30} chlorine atoms in the target. In an ordinary solution of chlorine compound (CCl_4) there are around 10^{23} chlorine atoms cm^{-3}. Thus if we use a tank of half a million litres of compound solution, we can expect a reaction rate of approximately a few times a day.

Another question is how to prevent the same interaction from occurring by some other energetic radiation that reaches the Earth from space. This problem can be overcome by locating the detector deep underground so that the thick layers of ground above absorb any radiation other than neutrinos.

Davis[2] constructed such a detector by placing a tank of 400,000 litres filled with chlorine solution in a mine at Homestake, Dakota, at a depth of 1600 metres. He extracted the argon produced from the solution. The argon is unstable, and after a lifetime on the order of days, it decays back to chlorine by emitting a positron and a neutrino. The number of atoms produced are counted through the inverse reaction:

$$^{37}Ar \longrightarrow \, ^{37}Cl + e^+ + \nu. \tag{12.2}$$

A standard unit for a neutrino count used in comparing observations by different detectors is the *Standard Neutrino Unit,* "SNU," where 1 SNU = 10^{-36} neutrino absorptions sec^{-1} per target atom.

The expected neutrino count calculated recently[3] yielded an expected rate of about 7.9 SNU for the chlorine detector. The actual results found by Davis detector are approximately 2.1 SNU, which are below the expected results by a factor greater than three. These results have been consistent for the preceding fifteen years during the use of this detector. A neutrino count carried out from 1987 to 1988 found a rate of 4.5 SNU,[4] which is about half the expected rate. Several theories tried to explain the discrepancy between the expected results and the observations.

One line of explanation proposes that energy production in the Sun is not a smooth process. Instead it involves ups and downs, wherein the observed solar luminosity is some average over these ripples. According to this explanation, we are now watching one of the downs in the Sun's energy production, which also results in a decrease in neutrino production. Some scientists suggest that a contraction of the Sun that releases gravitational energy replenishes the energy deficit during the downs in production. However no other observed signs can establish this theory.

An alternative suggestion is that the lifetime of neutrinos is short. On their way from the Sun, part of them decay, and only part of the original flux reaches the Earth. However, the fact that a significant flux of neutrinos created by SN 1987A reached the Earth from a distance of 165,000 LY puts this suggestion into serious question.

Another line of explanation relies on the suggestion that if for some reason the central temperature of the Sun is lower than that of the standard model, then the relative weight of energy production by PPI channel increases, and the weight of the energetic neutrinos produced in the nuclear reactions declines.

There are several possible explanations for the cause of this decrease in the central temperature of the Sun. One suggestion is that the central part of the Sun undergoes continuous mixing by some kind of convection. This model is called the *mixed model*, according to which the abundance of hydrogen at the solar centre remains high due to the mixing and requires a lower temperature to retain the same level of energy production. However the physical reasons for developing such convection are unclear.

According to another line of thinking, the reduced temperature at the solar centre is due to the presence of nonbaryonic particles inside the Sun that induce very high heat conduction locally. This high heat conductivity eases the demand on the temperature gradient and makes a lower temperature at the centre possible. This process occurs at the expense of a steeper density gradient forming the pressure gradient needed to support the Sun against gravity. These nonbaryonic particles are called *weakly interacting massive particles* (WIMPs), and their proposed mass is a few times the proton mass.[5]

According to a more extreme solution of this kind[6] the WIMPs emit a dark radiation in addition to having the role of enhancing heat conductivity. This dark radiation does not interact with matter, and energy propagates directly from the solar interior outward without being absorbed on the way. The WIMPs thus form an energy sink at the solar centre. Such particles are evidently very efficient in reducing the central temperature of the Sun.

As a result of this heat conductivity, the temperature gradient is very moderate at the solar centre. Moreover, the region in which nuclear reactions occur is broader, and the production of high energy neutrinos slows while that of low energy neutrinos increases. One of the advantages that the WIMP hypothesis offers is that their existence may solve another problem — that of the "missing mass" in the Galaxy. Dynamic calculations of the balance between the centrifugal force and gravity in the Galaxy show that the observed mass in the Galaxy is insufficient to supply the demanded gravitational balance. In fact, the hypothesis of the existence of WIMPs was proposed first to solve the problem of the missing mass in the Galaxy, and only later did researchers adopt it to solve the solar neutrino problem.

To date, however, no evidence has been found to substantiate the existence of such particles.

We have already noted that the chlorine-based detector can only detect high energy neutrinos, which account for less than 10 percent of the neutrinos produced in the Sun. We might obtain a clearer picture if we were able to detect low energy neutrinos, which make up the majority of the neutrinos produced in the Sun. Another kind of detector fulfills this purpose, in which gallium (^{71}Ga) is the detecting matter.

In natural gallium, 39.6 percent is ^{71}Ga, which upon interacting with a neutrino emits an electron and converts to germanium:

$$^{71}\text{Ga} + \nu \longrightarrow {}^{71}\text{Ge} + e^-. \tag{12.3}$$

Here the energy difference between the ground states of the two elements is 0.233 MeV. Thus it can detect even low energy neutrinos produced by the PPI channel. ^{71}Ge is an unstable isotope with a lifetime of 11.4 days which decays back into gallium. The detector spots this decay and count the events of neutrino absorption.

Two such detectors exist and are even now expected to attain a working state. One is the laboratory called GALLEX, in Gran-Sasso, Italy, built in a tunnel excavated under the Alps, parallel to a new highway. GALLEX contains a tank designed to hold 30 tons of liquid gallium. Researchers will extract the germanium created in its tank from time to time and inspect it for radioactive germanium atoms.

The second laboratory is called SAGE, and is in Baxan, under the Elbrus mountain, in the Caucasus. It contains 60 tons of metallic gallium, of which 39.6 percent is ^{71}Ga.

Table 12.2 also gives the cross-section for the different classes of neutrinos with gallium. Calculating the values obtained by multiplying these cross-sections by the corresponding neutrino fluxes at the Earth's surface yields the expected neutrino count by the gallium detector derived from the standard solar model as about 132 SNU. Undoubtedly, discrepancies in such a value (if discovered) are more significant than those found by the chlorine detector. The detection of the neutrinos with the gallium detector may settle, at least in part, the debate over the solar neutrino problem.

It is interesting to note that the solution to the solar neutrino problem of using the dark radiation by WIMPs demands a higher production of nuclear energy. It has to cover the energy loss by the dark radiation in addition to the solar luminosity observed regularly. Thus, the neutrino count expected

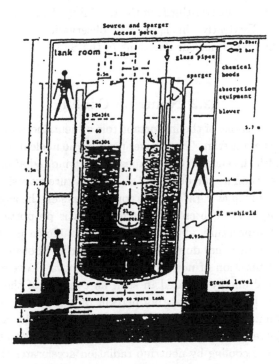

Figure 12.1. Schematic side view of the GALLEX Detector Tank with 30 tons of gallium. [Adopted from Barish.[7]]

from the gallium detector in the model with dark radiation is higher by 20 percent than the value expected from the standard model. Clearly, an observation of such a high count by the gallium detector would support this hypothesis.

In Chapter 10 we mentioned the Kamiokanda water detector for neutrinos. This detector, originally designed to detect proton decay, converts to a neutrino detector. By lowering its threshold for detected energies to about 5 MeV, it becomes capable of detecting the ^{8}B neutrinos from the Sun. The third line of tab. 12.2 gives the cross-section for this detection. Over a period of 450 days, it finds a counting rate equivalent to 4.5 SNU, thereby supporting the recent findings by Davis.

Recently the GALLEX team in Gran-Sasso completed nearly a year of observations and announced[8] detection of about 83 SNU of solar neutrinos. This amounts to about two-thirds of the total flux of 132 SNU predicted for solar neutrinos. Evidently, this is the first observation of neutrinos created by the PPI channel of nuclear reactions in the Sun.

12.2 Neutrino Cooling

In Chapter 6, we spoke of the important role of neutrino radiation from the hot cores of red giant stars. These cores produce no nuclear energy. The core cannot replenish the energy that escapes by neutrino radiation, so its centre becomes cooler than its surroundings. A temperature inversion forms which can reach to as much as $0.4 M_\odot$ in the cores of stars on the asymptotic giant branch (see fig. 4.7). The creation of neutrinos is proportional to a high power of the temperature, about T^6 to T^8. With increasing temperatures, neutrino cooling becomes dominant over cooling by photon radiation.

The cooling function of neutrinos in a supernova II event is more dramatic. The total energy released in such an event is of the order of 10^{53} erg. Neutrinos, however, radiate 99 percent of this energy away. Only an amount of about 10^{51} erg is observed as stellar luminosity and kinetic energy of the ejected material. When the collapse that leads to a supernova II event begins, the cooling by neutrino radiation accelerates the collapse.

The creation of neutrinos in hot matter proceeds through a number of channels: (1) the annihilation of electron-positron pairs; (2) the disintegration of plasmons to neutrino-antineutrino pairs; (3) photoneutrino process; (4) Bremsstrahlung radiation.

Annihilation of Electron-Positron Pairs

The annihilation process occurs in the form:

$$e^- + e^+ \Rightarrow \nu + \bar{\nu} \tag{12.4}$$

where $\bar{\nu}$ (antineutrino) is the antiparticle of the neutrino ν. High energy photons created the electron-positron pairs participating in these reactions. Such a creation of pairs is possible when the photon energy is above the rest energy of an electron-positron pair; in other words, if it is above one MeV. This process becomes efficient at temperatures of $\frac{1\text{MeV}}{k} \simeq 10^{10}$ degrees, where k is Boltzmann constant.

Disintegration of Plasmons

Here, a plasmon disintegrates to a neutrino-antineutrino pair:

$$\text{Plasmon} \Rightarrow \nu + \bar{\nu}. \qquad (12.5)$$

A plasmon is a quantized electromagnetic wave traveling in a dense dielectric plasma. Its energy is $\hbar\omega_p$, where ω_p is the plasma frequency.

Photoneutrino Process

This is a process in which an energetic photon interacts with a free electron to form a neutrino-antineutrino pair:

$$\gamma + e^- \Rightarrow e^- + \nu + \bar{\nu}. \qquad (12.6)$$

Bremsstrahlung Radiation

Here neutrinos are created through Bremsstrahlung radiation (braking radiation), when a high energy electron collides with a heavy nucleus:

$$e^- + (Z, A) \Rightarrow (Z, A) + e^- + \nu + \bar{\nu}. \qquad (12.7)$$

The energy of the neutrinos created in the above processes is of the order of kT, where T is the temperature of the system. These processes continue to cool down the neutron star formed in the SN II for as long as its central temperature is sufficiently high. Only when the central temperature decreases to below 10^8 degrees does cooling by radiation of photons become more efficient than neutrino cooling. An important process in neutrino cooling is called URCA process. In this process, which takes place at very high temperatures, the creation of an antineutrino in the decay of a neutron to form a proton is followed by the inverse process, which recovers the initial state through an electron capture by the proton creating another neutrino:

$$\begin{aligned} n &\Rightarrow p + e^- + \bar{\nu} \\ e^- + p &\Rightarrow n + \nu \,. \end{aligned} \qquad (12.8)$$

Overall the two stages recover the neutron but also create a neutrino-antineutrino pair which drains energy from the system. The name URCA comes from a well known gambling casino in Rio de Janeiro, where after all manipulations of the odds the players are finally drained of all their cash. The URCA process is efficient for free particles only at high temperatures of above 10^{10} degrees, when the kinetic energy of the electrons is on the order of their rest energies. When the temperature falls below this value, the conservation rules of energy and momentum cannot be satisfied simultaneously for free particles. The process can proceed only in the presence of another heavy particle capable of absorbing the excess momentum:

$$
\begin{aligned}
n + n &\Rightarrow n + p + e^- + \bar{\nu} \\
n + p + e^- &\Rightarrow n + n + \nu .
\end{aligned}
\tag{12.9}
$$

This process is called the *modified URCA process*. The calculations show that the luminosity of URCA neutrinos, L_ν, of a star with uniform density is given by:

$$
L_\nu^{URCA} = 5.3 \times 10^{-33} T^9 \, M \left(\frac{\rho}{\rho_{nuc}} \right)^{-1/3} \quad \text{erg sec}^{-1}
\tag{12.10}
$$

where M is given in solar units and ρ_{nuc} is the nuclear density $= 2.8 \times 10^{14}$ gm cm^{-3}. It is clear that URCA neutrino luminosity reaches values on the order of solar luminosity only at temperatures around 10^8 degrees. More details concerning this topic may be found in Shapiro and Teukolsky.[9]

A more general form of an URCA process is one in which the neutron decay and the electron capture by the proton take place inside a nucleus:

$$
\begin{aligned}
(Z, A) &\Rightarrow (Z + 1, A) + e^- + \bar{\nu} \\
e^- + (Z + 1, A) &\Rightarrow (Z, A) + \nu .
\end{aligned}
\tag{12.11}
$$

This reaction takes place under conditions of high electron degeneracy. The relevant electrons for the electron capture are those at the highest energy level, which is the Fermi energy, ϵ_F, of the system. The electron capture is an endoergic process, and the reactions can proceed if ϵ_F equals the threshold energy for the electron capture, ΔQ is different for different species of elements. This condition, $\epsilon_F = \Delta Q$, is valid at different temperatures for each pair of elements (Z, A), $(Z + 1, A)$, which are called URCA pairs. The region which satisfies this condition for a particular pair is called the URCA shell for that pair of elements.

In a convective region where a matter cell travels along a certain temperature gradient, this matter cell passes through consecutive URCA shells of different URCA pairs. The URCA process may take place at each URCA shell in turn. However, when the URCA process occurs in a convective region it causes local cooling which disturbs the initial temperature gradient, and these disturbances may change the initial convective regime. Overall, such an interaction between the URCA process and convection may support fast cooling and prevent explosive events when they are otherwise likely to occur.

The cooling by URCA process in a convective zone has an implication for the evolution of medium mass stars. At the AGB phase a star has a carbon-oxygen (CO) core, two burning shells (helium and hydrogen), and a hydrogen-rich envelope. In high mass stars (above $8M_\odot$) the central temperature is high, and the matter is nondegenerate. As the central temperature reaches the threshold for carbon ignition, further nuclear burning takes place and the star proceeds toward a supernova II event (see Chapter 10). However in stars with masses below $8M_\odot$ having lower central temperatures, the matter in the core is degenerate, and the carbon ignites explosively.

The fate of the star depends, to a large extent, on the efficiency or inefficiency of the cooling mechanisms that govern the process. Heat transfer by radiation or convection is much too slow to carry away the huge amounts of energy released in the explosion. The result of detailed calculations of the explosion process is that the explosion disrupts the entire star. This will also occur in an SN I event. If some mechanism exists to cool the system significantly, this may moderate the explosion in such a way that a regular SN II event may proceed, producing a remnant in a form of a neutron star. A convectively driven URCA process in the exploding core may supply an efficient cooling process which will moderate the explosion so as to become an SN II event.

Research has studied[10] this effect. The calculations are very complicated because the interaction between the URCA process and convection must be computed in detail at each point. The crucial parameter is the time scale of the convection, τ_c, inferred from the convection velocity. We compare this time scale to the time scale of the URCA process, τ_U. In order to produce an effective cooling, τ_c must be very small compared to τ_U, namely $\frac{\tau_c}{\tau_U} \ll 1$. Calculations of the energy released in the explosive carbon ignition yield a nuclear luminosity of about $5 \times 10^6 L_\odot$. The neutrino cooling by convectively driven URCA process yields a cooling luminosity of approximately $5 \times 10^5 L_\odot$, which is marginally sufficient to moderate the explosion and prevent disruption of the star. We need more detailed calculations performed by improved methods and updated input physics to arrive at a definite answer as to whether a star that possesses four solar masses at the phase of carbon ignition will end this ignition as an SN I or SN II event.

We also consider the cooling effect by the URCA process in the evolution of neutron stars, a topic with which we deal in the following section.

12.3 Neutron Stars

In Chapter 6 we described how a white dwarf forms from the bare core of a red giant star. No further nuclear reactions take place in this star, which cools down as it loses energy by radiation. Thermal pressure decreases with the decrease in temperature, and the star proceeds to contract. With increasing density the matter becomes highly degenerate, and the pressure created by the degenerate electron gas supports the star against gravitational collapse. A study of the polytropic configurations (presented in Chapter 4) shows that there is a limit to the stellar mass which a degenerate electron gas can support. This is the Chandrasekhar limit and is about $1.44 M_\odot$. What is the fate of a cooling star whose mass is above this limit?

Calculations of the structure of white dwarfs show that they attain the Chandrasekhar limiting mass at a central density of 10^9 to 10^{10} gm cm^{-3}. Cool stars having a mass higher than the limiting mass will contract, and their central density will increase further. Up to densities of 10^7 gm cm^{-3}, the baryonic matter is organized in iron atoms, $^{56}_{26}$Fe, which is the lowest energy atomic state. When the density exceeds 10^7 gm cm^{-3} the electrons become relativistic, and their kinetic energies are sufficiently high to interact with protons, to form neutrons by emitting an antineutrino:

$$e^- + p \rightarrow n + \bar{\nu}. \qquad (12.12)$$

This interaction is called the *inverse β decay* because it is an inverse process to regular β decay, in which a neutron decays to a proton and an electron. (Note that a sequence of inverse β decay followed by β decay is what forms the URCA process which the preceding section, eq. 12.8, describes.) When inverse β decay takes place inside a nucleus of an atom, the result is a neutron-rich matter, and the ratio n/p in the matter increases. The neutronization of matter creates larger nuclei. Upon reaching a density of about 4×10^{11} gm cm^{-3}, a *neutron drip* takes place in which free neutrons form, and the matter exists in a two phase system, one of electrons and nuclei and the other of neutrons. The fraction of free neutrons increases with further increase of density, and they contribute their partial pressure to the total pressure in the system. When the density increases to above 4×10^{12} gm cm^{-3}, the pressure contributed by free neutrons dominates the system.

The density, ρ, and the pressure in the system are at this stage given by:

$$\rho c^2 = \epsilon = n_e \frac{M(A, Z)}{Z} + \epsilon_e + \epsilon_n$$

$$P = P_e + P_n \qquad (12.13)$$

where P_e and P_n are the partial pressure contributed by electrons and neutrons respectively. ϵ is the total energy per unit volume, which includes also the rest energies of the particles. ϵ_e, ϵ_n are the energies per unit volume of the electrons and neutrons, respectively. n_e/Z is the number density of nuclei and the first term on the right-hand side of the first line of eq. 12.13 represents the energy of the nuclei. The contribution of the nuclei to the pressure is negligible relative to that of the degenerate electrons and free neutrons. At these high densities, electron degeneracy is very high, and we can ignore the contribution of the temperature to the pressure. The relation between the pressure and the density is actually the equation of state of the system at the given conditions.

With a given equation of state, we can use the hydrostatic equation to construct a model of a star in a hydrostatic equilibrium. However the hydrostatic equation cannot be used in its Newtonian form (eq. 2.2) since, owing to the high density, we cannot ignore general relativistic effects. We instead use the relativistic hydrostatic equation derived from Einstein's equations for the gravitational field. This equation is given by:

$$\frac{dP}{dr} = -\frac{G(m + \frac{4\pi r^3 P}{c^2})(\rho + \frac{P}{c^2})}{r^2(1 - \frac{2Gm}{c^2 r})}. \tag{12.14}$$

Inspection of this equation shows that when $P \ll \rho c^2$, and $\frac{2Gm}{c^2} \ll r$, the Newtonian equation recovers from eq. 12.14. Let us analyze the separate parts of eq. 12.14. First we note the double appearance of the pressure in the right-hand side of the equation: in the first parenthesis the pressure is added to the mass, thereby enhancing the source of gravity in the system; and in the second parenthesis it is added to the density, thereby enhancing the quantity upon which gravity acts. In both cases we find that the pressure increases the gravitational force in the system. But since it appears as P/c^2, it is clear that such a term becomes significant only at very high pressure. On the left-hand side of the equation, an increase in the pressure leads to an outward force through the steepening of the pressure gradient, whereas on the right-hand side the same increase in the pressure also increases the gravitational force which drives a contraction.

This form of the equation manifests the gravity of the kinetic energy, which is proportional to the pressure. It demonstrates that highly dense systems favour collapse. The increasing pressure, usually expected to contradict a collapse, now drives the collapse. With further increase of density and pressure, the importance of gravity deriving from the pressure increases as well.

Another correction appears in the denominator of the right-hand side of eq. 12.14: the term $\frac{2Gm}{rc^2}$ in the parenthesis. Obviously when this term is negligible, the usual denominator of r^2 appears in the equation. However when the same term approaches unity, it causes the right-hand side of eq. 12.14 to increase indefinitely, and the demand on the pressure gradient increases accordingly. This term represents the role of the specific radius: $r_S = \frac{2Gm}{c^2}$, where r_S stands for Schwarzschild's radius. When this equality is valid, the right-hand side of eq. 12.14 diverges, and even an infinitely steep pressure gradient cannot prevent a collapse and the star will turn into a black hole. The surface defined by this value of the radius is the event horizon of a Schwarzschild black hole from which no matter, not even photons, can escape. More details about black holes, event horizons, and related phenomena can be found in textbooks on general relativity.

The equation of mass continuity:

$$\frac{dm}{dr} = 4\pi r^2 \rho \tag{12.15}$$

can be used together with eq. 12.14. Integrating the two equations, using the appropriate equation of state at each given point, yields the stellar configuration. The integration begins at the centre, with a chosen central density used as a parameter, and proceeds outward until reaching a surface at which the pressure (or density) vanishes. This surface is the boundary surface of the star. The mass accumulated through the integration of eq. 12.15 from the centre to the boundary surface is the stellar mass. Each choice of a central density, ρ_c, yields a certain value for the stellar mass enclosed in the configuration.

Figure 12.2 displays the values of the stellar masses calculated for each choice of ρ_c, where we plot M vs. ρ_c. These values are calculated by integrating eqs. 12.4 and 12.5. The equations of state used for each value of the density are given by Canuto.[11]

The behaviour of the graph in this figure shows the dependence of the stellar configuration on the character of the equation of state at each range of the central density. Starting with densities of 10^5 gm cm^{-3}, the stellar mass increases with increasing ρ_c, since increasing ρ_c means more compact packing of the mass. This behaviour continues, as might be expected, up to densities of about 10^{10} gm cm^{-3}. The mass reached at this ρ_c is around $1.2M_\odot$, which is close to the Chandrasekhar limiting mass. With further increase in the central density, the mass contained in the configuration decreases. This is due to the change in the equation of state deriving from the neutronization of matter which takes place at these densities.

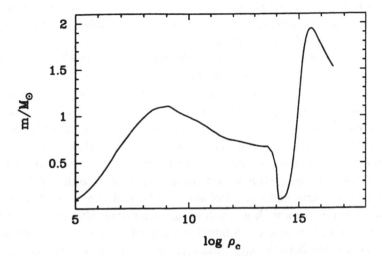

Figure 12.2. Masses of compact cool stars vs. ρ_c.

As long as the mass contained in the configuration increases along with the central density, the configuration is stable. An increase in the mass results in stronger gravity which drives contraction. The contraction results in higher density which enables accomodation of the increased mass. Thus the range in ρ_c for which the graph ascends represents stable stellar configurations. The situation is different in the range for which the graph descends ($10^{10} < \rho_c < 10^{14}$ gm cm^{-3}). An increase in mass, which results in contraction and an increase in central density, drives the configuration to a central density which cannot accomodate the increased mass. If the star can get rid of the excess mass it can remain within the region of stability. However no suitable mechanism for mass ejection exists in a cooling star. The response of the star is to contract further, thereby further increasing its central density and the range in which the stellar mass decreases with an increase of the central density represents unstable stellar configurations. If no change takes place in the form of the equation of state, this contraction will continue forever, more drastically for higher than for lower mass stars.

Before establishing the existence of neutronic matter, it was commonly believed that cooling stars having masses above the Chandrasekhar limit would contract indefinitely and form black holes. As a result of both observations of cool stars with masses above this limit, during the 1960s, and advances in our understanding the structure of matter at very high densities, the idea emerged of the existence of neutronic matter. The concept of neutron stars in which most of the matter is neutronic was established. Calculating the equation of state for such matter enabled us to calculate the configurations of neutron stars.

When the density reaches values of the nuclear density (2.8×10^{14} gm cm^{-3}), the behaviour of the equation of state changes. We observe in fig. 12.2 that the graph ascends again, showing the existence of stable solutions in that region of ρ_c. This region extends to values of ρ_c of few times 10^{15} gm cm^{-3}, and it enables stable configurations to exist with masses above the Chandrasekhar limit. Thus we find two branches of stable configurations for cool stars — the white dwarf branch and the neutron star branch. The accepted picture of a neutron star is of an object with a radius of 10 kilometres and a mass of about $1M_\odot$ and above. Neutron stars possess an outer crust with a depth of one kilometre. Beneath this crust the neutronic matter exists in a liquid form, starting with a density which equals the nuclear density, ρ_{nuc}, just beneath the crust and increasing inward.

The equation of state constructed from eqs. 12.13 suits the density range between the neutron drip density and the nuclear density, namely $\rho = 10^{11} - 2.8 \times 10^{14}$ gm cm^{-3}. In this range the matter consists of neutron-rich nuclei, free electrons, and free neutrons. When the density increases above the nuclear density, nuclei start dissolving and merging. The equation of state at these densities is not definitely known and is subject to many uncertainties. Each assumption made for a certain equation of state yields a different value for the mass calculated by the integration of eq. 12.14.

In fig. 12.3 we display a variety of solutions of M vs. ρ_c, with central densities above the nuclear density. The letters marking the graphs designate the specific equation of state used to calculate the models.

For a detailed treatment of this topic the reader may refer to the work of Baym and Pethick,[12] and to the references cited therein. For the sake of completeness only, we give the meaning of the letters marking the graphs in fig. 12.3:

MF — Mean field theory calculations.

TI — Tensor interaction model.

BJ — Bethe-Johnson equation of state, which includes small effects of hyperons.

R — Equation of state of pure neutronic matter, with Ried potential.

$\pi\ \pi'$ — Ried's equation of state of neutronic matter, with modifications owing to charged-pion condensations.

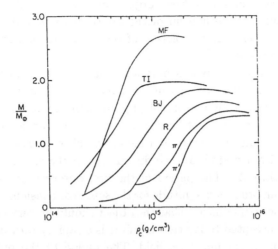

Figure 12.3. Models calculated with different equations of state. [Adopted from Baym and Pethick.[12]]

It is interesting to note the variety of masses that a stable model can attain with different equations of state. The range of maximum masses displayed in fig. 12.3 is 1.5 to 3 $M_\odot$. Further advances in the physics of highly dense matter may lead to a more definite value for the maximal mass available for a neutron star.

Understanding the properties of neutronic matter (supported by the observations of stellar objects with masses at and above the Chandrasekhar mass) pushes the limiting mass of a stable configuration that will not collapse to a black hole to a higher value. Further investigations of systems denser than neutronic matter, such as *quark matter*, may reveal a third branch of stable cold stars in the M/ρ_c diagram in addition to the white dwarf branch and the neutron star branch.

12.4 Pulsars

The discovery of the interesting objects called *pulsars* occured only in recent decades A British team at Cambridge University[13] found the first. The most prominent property of these objects is their radiation of short pulses in the radio wavelength (around the frequency of 400 MHz) having periods of around one second or less. They emit these pulses with a very exact timing.

Not all the observed details of these objects are well understood, and the theory of the emitting mechanism of the radio pulses is still incomplete. However a reasonably coherent picture of these objects emerges from the accumulated observations.[14]

The short period of the pulses puts a strong constraint on the dimensions of the radiating object. Causality demands that the dimensions of an object which creates a time-dependent phenomenon demanding a coherent behaviour of the source will not be greater than the distance traveled by a light signal during the characteristic time of the phenomenon. Thus for a periodic pulse with a period Δt, the dimensions of the source should not exceed the length, $c \cdot \Delta t$. The period of the first observed pulsar is 0.033 second, yielding an order of magnitude for the source dimensions as less than 10^8 cm. This observation means that the object should be very compact.

It is generally accepted today that a pulsar is a rapidly rotating neutron star possessing a strong magnetic field. The period of the pulses is the rotational period of the star. A directed beam, probably emitted in the direction of the axis of the magnetic field, produces the observed pulse. The magnetic axis and the rotation axis align with each other at a certain angle. Due to this alignment the directed beam corotates with the magnetic axis around the axis of rotation. During this rotation the directed beam sweeps the sky. From objects like the Earth that happen to reside on the sweeping circle of the beam, we observe the pulsation each time the beam crosses our path. Evidently the sweeping only covers a certain circle of the sky, whose width depends on the beam width, and only a part of the pulsars are observed from the Earth. The fraction of pulsars observed depends on the shape of the beam. If the form of the beam is a circle, then about 20 percent of the pulsars are visible from the Earth.

More than 500 pulsars have been observed in the Galaxy. One was observed in the Large Magellanic Cloud. The luminosity radiated in the beam is not high and is too weak to be observed from other galaxies.

Neutron stars form in SN II explosions or through mass accretion in close binary systems. However we have observed only three pulsars in supernovae remnants (SNRs). These pulsars are young pulsars. One resides in the Crab Nebula, which is the remnant of SN 1054, and it is less than a thousand years old. Another is the Vela X pulsar, whose age is roughly 12,000 years. A third is called G320.4-1.2 (1509-58) and is also a relatively young pulsar. Other pulsars are likely older with estimated ages of more than 6×10^4 yr.

We expect to see a nebula of SNR for about 10^4 yr, after which it fades into the interstellar medium. Since the estimated lifetime of pulsars is more than a hundred times longer than that of SNRs, it is reasonable to find one percent of the pulsars inside SNRs.

For part of the pulsars we see an increase in the pulse period, which means that stellar rotation is slowing down. From the rate of the increase of the period, we infer an estimate of the pulsar age. Denoting the pulse period by P and its time derivative by $\dot{P}$, the quantity $\frac{P}{\dot{P}}$ is a measure for the characteristic time, T_P, needed for the period to increase by a significant fraction of its initial value. For many pulsars this is:

$$T_P = \frac{P}{\dot{P}} \sim 10^7 \text{ yr.} \tag{12.16}$$

It is interesting that in the case of the three "young" pulsars mentioned above, the periods (0.033, 0.09, 0.15 seconds) are shorter than the average (which is 0.2 to 1 second). The values of T_P of the same three pulsars are smaller than the average. This suggests that the rate of increase in the period may be relatively higher for shorter periods — namely in pulsars that rotate faster. When the period increases with the aging of the pulsar, the rate of period increase decreases.

If pulse luminosity remained constant with age, we would expect to find more pulsars with longer periods. Observations show that for periods of above $P \sim 1$ sec, there is a steep decline in number of observed pulsars. This indicates that luminosity probably decreases with age, and disappears entirely in periods of above four seconds.

Another measure of the age of pulsars is their location above the Galactic disk. Half the observed pulsars exist at a distance of less than 800 LY above the Galactic disk. Pulsars probably form from high mass stars in the Galactic plane. They have high velocities moving away from the Galactic disk which are of the order of 100 km sec^{-1}, whereas ordinary disk population stars have velocities of 30 to 40 km sec^{-1}. We think that the pulsars acquired their high velocities when they were created in the SN explosion. Their distance from the disk is proportional to their velocity times their lifetime after the explosion. If we denote their distance above the disk by Z and their velocity in this direction by v_z, we have for their lifetime, T_{life}:

$$T_{life} = \frac{Z}{v_z} \simeq 10^7 \text{ yr} \tag{12.17}$$

which well suits the value found for T_P in eq. 12.16.

The 500 pulsars found thus far are located around the Sun, at up to a distance of 6,000 to 8,000 LY. From the average number of pulsars per unit volume calculated from these observations, we estimate a number of about 10^5 active pulsars for the entire Galaxy.

The mechanism which produces the radio pulses is not yet clear. As observed by Rankin,[15] the radiated beam consists of two components: a core and a hollow cone, with differing characteristics. There are subpulses in the radiation of the hollow cone that may drift forward or backward relative to the main pulse. Some pulsars show nulls in their pulse radiation; these are cessations in their pulse radiation for the duration of several periods. We find that pulsars having only the hollow cone in their beam do not have nulls in their radiation.[14] On the average the beam core radiation accounts for 60 to 70 percent of the total pulse radiation. It may be that the two components of the beam form at different locations on the star.

When charged particles accelerate, radio radiation forms. A star possessing a magnetic field creates a magnetosphere around its body containing charged particles whose motion is locked along the magnetic field lines.

The field lines in the magnetosphere emerge from a region around one pole and terminate in a region around the other pole, completing a closed loop along the axis of the magnetic field. The part of the loop outside the star is high above the equator. When the star rotates, the magnetosphere rotates with it, and all the matter particles that are locked to the field lines in the magnetosphere participate in the rotation. However at a certain distance from the axis of rotation, the radius of rotation, r_c, reaches a value for which $\omega \cdot r_c = c$, where ω is the angular velocity of the rotation. The velocity of the corotating matter at this radius approaches the speed of light. Evidently, the corotation breaks down at r_c. The cylinder described by r_c is called the light cylinder. Matter particles can travel along the field lines in the magnetosphere from pole to pole, where the light cylinder encloses the loop of the field lines. However, magnetic field lines that emerge in close vicinity to the poles must complete their loop outside the light cylinder, and matter particles cannot travel along these field lines. The *polar cap* is the region around the pole from which field lines that do not complete their loop inside the light cylinder emerge. Particles moving along these field lines are subject to very high acceleration, which causes the radio radiation.

Part of the acceleration is curvature acceleration: due to the small dimensions of the star and highly dense magnetic flux, the curvature of the magnetic field lines near the poles is very strong. The curving of the particles' trajectories along the curvature of the field causes a transverse acceleration which gives rise to a part of the radiation. A longitudinal acceleration around the field lines creates the other part of the radiation.

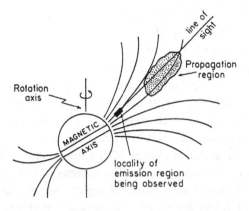

Figure 12.4. Geometry of a pulsar system. [Adopted from Rankin.[15]]

Radiation of the beam core forms very close to the stellar surface at the polar cap. Rankin[15] calculated the opening angle of the beam core, W_{core}, and showed that it depends on the period duration, P, and on the angle α between the rotation axis and the axis of the magnetic field:

$$W_{core} = \frac{2.45^o}{P^{1/2}\sin\alpha} \qquad (12.18)$$

where P is given in seconds. For most of the pulsars, α is around 35 degrees. Figure 12.4 shows the geometry of the pulsar and the relevant angles.

The radiation of the hollow cone forms at higher latitudes above the polar cap than the origin of the core radiation. Its opening angle is wider than that of the beam core.

Twelve of the pulsars are found in binary systems. This observation is surprising since we know that in an ordinary stellar population more than 50 percent of the stars reside in binary systems, and we would expect to find the same percentage in the pulsar population. A reasonable explanation of this observation might be that since most pulsars form as a result of a supernova explosion, the companion was kicked out of the system or, alternatively, engulfed by the primary to form the neutron star.

Researchers discovered three millisecond (ms) pulsars from 1982 to 1984. These are very fast rotators with pulse periods of 0.0015, 0.005, and 0.006 second. Their magnetic fields are relatively weak, and they are more similar to old pulsars, which generally rotate much more slowly. We believe that these are "recycled" pulsars which after aging and slowing down were spun up again through mass accretion from a companion in a close binary system. Indeed two of them are members of binary systems wherein the masses of the companion are 0.2 to 0.4 $M_\odot$.

One of the ms pulsars is a single star. Probably its companion is accreted entirely to the pulsar, or it evaporated by the pulsar radiation after recycling the rapidly rotating pulsar. This pulsar is named a "black widow," after the female spider that feeds upon the male after mating[16].

A number of ms pulsars were observed in globular clusters that are sufficiently close to enable their observation. For instance, eleven ms pulsars are observed in the globular cluster 47 Tucanae, whose distance from us is about 13,000 LY. It may be that the increased abundance of ms pulsars in globular clusters results from the higher rate of mutual interactions between the stars in the dense stellar population at their centre.

Recently, researchers observed a binary system consisting of a pulsar and a Be-star.[17] The pulsar is designated as PSR 1259-63, and the Be companion as SS 2883. The pulse period of this pulsar is 47.76 milliseconds, and the orbital period of the system is about 2,000 days. We identify the companion as a Be-star (recall Chapter 9) by the strong emission lines found in its spectrum and by the rapid rotation indicated by the Doppler broadening of the spectral lines. The mass of the companion is above $12M_\odot$. The orbit of the system has a very high eccentricity of about 0.967, which corresponds to a ratio of 0.25 of the minor axis to the major axis of the elliptical orbit. A clear eclipse occurs in the pulse sequence, which appears at the periastron of the orbit and lasts for about twenty days. The eclipse terminates gradually; meaning that it probably results not only from the star itself but also from the dense wind around the star.

The distance to the system from Earth is 7,500 LY, and the luminosity of the Be-star is $5.8 \times 10^4 L_\odot$. As described in Chapter 9, in the further evolution of this system the Be-star will explode through a supernova event, and if the explosion does not disrupt the system, a binary of two neutron stars will form.

12.5 SS 433

The object SS 433 was the target of very intensive observations during the last decade. Its designation indicates that it has the ordinal number 433 in the 1977 catalog prepared by Stephenson and Sanduleak (SS). This catalog lists stars showing strong emission lines in their spectra. The object's location inside the supernova remnant (SNR) W50 suggested that it might be connected with the neutron star left after the SN explosion which created SNR W50. Scientists also observed it in the radio spectrum and in X-rays, so they expected to find an interesting object there. Initially, the different observations yielded different names in each range of the spectrum — 4C04.66 in the radio spectrum, A1909+04 in X-rays, and SS 433 in optical observation. It was finally established that all these different observations refer to one and the same object.[18]

The most peculiar observation of this star is the finding of very high Doppler shifts in its optical spectrum, in both directions — blue shift and red shift simultaneously. Moreover the shifted lines found in the spectrum were not constant but moving, their shift increasing and decreasing and even interchanging places in the spectrum for a period of 164 days. These observations demanded a careful analysis. In 1979 Milgrom[19] and Fabian and Rees[20] suggested a model known as the *kinematic model*. The kinematic model contains two collimated fast matter jets, ejected in opposite directions from a central object. These jets are the source of the high red and blue spectral shifts. The variation of these shifts with a period of 164 days is due to a precession of the axis of the jets, causing the jets to form a cone in the sky.

Analysis of further detailed observations revealed more details about the system:[21] the maximum velocity of the matter in the jets, inferred from the spectral shifts, is 50,000 km sec^{-1} in the red shift and 30,000 km sec^{-1} in the blue shift. The jets are highly collimated, with an opening angle of few degrees. The axis of the cone of the precession of the jets aligns at 79 degrees to our line of sight, and the half opening angle of the cone in the sky by the precession is 20 degrees.

The existence of the precession, and the need for a source of the matter ejected in the jets, demand a binary interaction. Indeed, researchers established the binary character of the system arriving at a binary period of 13.087 days. Taking the average between the velocities of the red and the blue shifting jets yields an observed velocity of 40,000 km sec^{-1}. Taking into account that we observe the projection of the actual velocities on our line of sight, the velocity of the jets at their centre of mass system results is 78,000 km sec^{-1} which is 0.26c, where c is the light velocity. The question is why should the red shift be larger than the blue shift. The system moves with a velocity of 70 km sec^{-1} away from us, but this motion does not explain why the red shift is greater than the blue shift by a factor of 1.6. The reason for this difference is that due to the high velocities, the spectral shift requires relativistic calculations. The spectral shift includes a transverse Doppler shift in addition to the ordinary longitudinal shift. The formula for the transverse shift is:

$$\lambda = \lambda_0 \frac{1}{\sqrt{1 - \frac{v^2}{c^2}}} \tag{12.19}$$

where λ and λ_0 are the observed wavelength and the source wavelength in its rest system, respectively. This calculation adds another red shift on top of the ordinary longitudinal shift, so that both shifts push toward the red end of the spectrum. The distance to the system is 16,000 LY, and the optical luminosity of the normal star in the system is $5 \times 10^4 L_\odot$, which puts it among the most luminous stars in the Galaxy.

We estimate the age of SNR W50 to be a few thousand years, and the age of the compact object of SS 433 is about the same. The kinematic model nicely explains the observations concerning the spectral shifts and periodic phenomena in the system. It does not explain the structure of the stars in the system, the mechanism of the jets' ejection, the collimation of the jets, and the evolutionary track that created such a system.

In fig. 12.5 we display a graph of the Doppler shifts[22] for the system calculated from the kinematic model for an epoch of 10 years. It also includes observations of these shifts during the 10 years from 1978 to 1988. The agreement between the two is quite good and shows that basically the kinematic model is a reasonable model for the system.

Superimposed on the two main periods — the 164-day period of the precession, and the 13-day period of the binary orbital rotation — there are additional small modulations with periods of 6.28 days and 5.8 days. It is proposed that these modulations are nods of the compact object induced on the accretion disk by the companion. The fact that the jets follow these nods implies that the disk affects (or causes) the collimation of the jets.

SNR W50, in which SS 433 resides, is an elliptical nebula whose major axis aligns in the direction of the axis of the jets with a length of 200 LY. The system of SS 433 is at the centre of the ellipse. We believe that the ellipse formed from a spherically symmetric nebula by the jets which pushed the nebula matter along their motion. Calculating the work invested in forming the ellipse yields, for the kinetic energy of the jets, a rate of $> 10^{39}$ erg sec^{-1}. The source of the kinetic energy for the jets is probably a mass accretion from a companion.

Two questions concerning the jets must be answered: what causes the exact high velocity of the jets with $v = 0.26c$; and what is the cause for the narrow collimation of the jets.

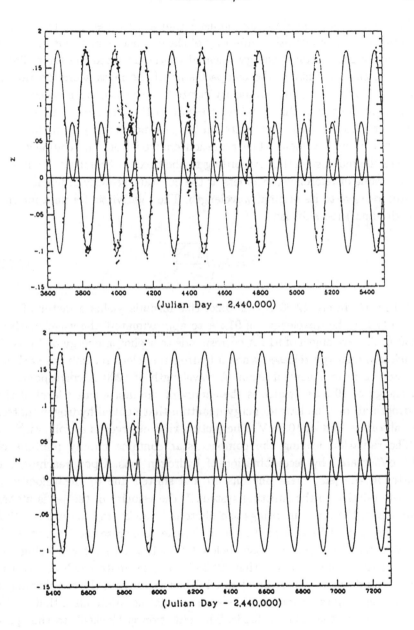

Figure 12.5. Doppler shift from SS 433. The solid lines display the shifts as calculated from the kinematic model. The dots display observations taken during this period. [Adopted from Margon and Anderson.[22]]

We believe that one of the mechanisms that accelerates the jets is radiation pressure. The intrinsic luminosity of the compact star is not high, but part of the gravitational energy released by the accreted matter radiates from the close vicinity of the compact star. If the mass accretion rate is above $10^{-8} M_{\odot}$ yr^{-1}, the luminosity created by this accretion exceeds the Eddington limit, and the radiation pressure of this luminosity accelerates mass particles outward. The matter in the jets moves away from the compact star at high velocity. The radiation emitted from the compact star reaches the moving matter. According to the theory of relativity, it moves in the frame of reference of the jets at the constant velocity c, but it is red shifted relative to its original wavelength. The relativistic formula for this (longitudinal) red shift is:

$$\lambda = \lambda_0 \sqrt{\frac{1 + \frac{v}{c}}{1 - \frac{v}{c}}}. \tag{12.20}$$

When v/c in eq. 12.20 equals 0.26, this formula yields a factor of 1.3, which brings the wavelength of 912 Å to approximately the wavelength of 1200 Å. A wavelength of 912 Å corresponds to a photon energy of 13.6 eV, which is the energy released when a hydrogen nucleus recombines with an electron to form a neutral atom. A wavelength of 1200 Å corresponds to an energy of 10.2 eV, which is the energy of the first excited level of the hydrogen atom. The lowest energy quantum that neutral hydrogen matter can absorb is that of 10.2 eV. The picture that emerges is as follows:[23]

The radiation leaving the compact star contains mainly photons of 13.6 eV created by recombination of hydrogen ions. Upon reaching the matter in the jets, these photons are red shifted due to the velocity of the jets relative to the emitting system. If the velocity of the jet is higher than 0.26c, the red shift brings the photons to an energy lower than 10.2 eV. They cannot be absorbed by and likewise accelerate the jet matter. If the velocity of the jet decreases below 0.26c, the red shift will bring the photons to an energy higher than 10.2 eV. The jet matter will absorb the photons and the jet will be accelerated. Thus we see that the red shifting of the accelerating radiation contains a self-regulating mechanism that limits the velocity to the value of 0.26c. The radiation is "locked" to the spectral line which represents the energy difference between the ground state and the first excited state of an hydrogen atom. Hence this mechanism is called the *line locking mechanism*. The jets brake by interacting with the interstellar matter. With the combination of this braking and the limited acceleration by the radiation, the jets maintain a constant velocity of 0.26c. It may be that radiation pressure is not the only accelerating mechanism, but it is undoubtedly the one which stabilizes the velocity of the jets at the above-mentioned value.

The accelerating radiation does not contain photons with energies higher than 13.6 eV because on leaving the compact star it passes through the cool gas of the disk. This gas "blankets" the radiation from higher energy photons, probably by absorbing the energetic photons.

This point brings us to the question of the narrow jet collimation. The collimation likely formed from a very thick accretion disk which covers the compact star almost entirely, leaving only narrow circular gaps at the poles. From these gaps the jets emerge. This structure somewhat resembles the bipolar flows of Herbig-Haro objects (see Chapter 8), where the emerging flows start at the poles, where the covering cloud of the infalling matter is the least dense. In a similar situation of SS 433, the matter accreted from the secondary star forms an accretion disk before losing its angular momentum and joining the compact star. The disk is thicker at the equator, where the particles possessing high angular momentum aggregate, and becomes thinner close to the poles, where the radius of rotation is much smaller and particles with very low angular momentum reside. The important question is how thick is the disk.

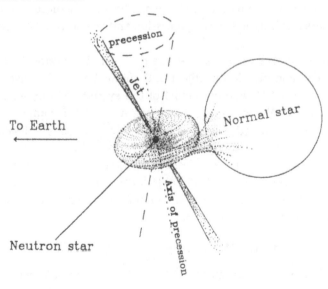

Figure 12.6. An artistic view and the geometry of SS 433.

Conclusions from analysis of the light curve of the star would indicate that the disk is remarkably thick. We suggest that the ratio of the height of the disk (above the equatorial plane) to its radius is two-thirds. The estimated radius of the disk is $\sim 10^{12}$ cm $\simeq 14 R_{\odot}$, which is of the order of magnitude of the separation between the binary components. Hydrodynamic calculations of the precession and short period nods observed in the precession also support the suggestion that the disk is thick and massive and contains high viscosity matter. We believe that this structure and the properties of the disk are the cause of the precession. If the rotation axis of the compact star, which is also the rotation axis of the disk, aligns at a certain angle to the orbital plane of the system, the secondary star will exert a torque on the disk matter whose edge is very close to the secondary. Due to the thickness and high viscosity of the disk, it responds almost like a solid body with a precession whose period of 164 days results from the magnitude of the torque and the properties of the disk. The orbital motion of the secondary, having a 13-day period, adds modulations to the precession which are manifested by the short period nods.

The dimensions inferred for the disk, with a height of a few orders of magnitude greater than the radius of the compact star, yields a picture of a "loafy" disk almost covering the compact star. The polar gaps from which the jets emerge are narrow and very deep, and this structure provides the narrow collimation of the jets. The same structure also explains why the jets so nicely follow the precession and the nods in the precession of the disk.

Intensive research in the last decade revealed many important details of this peculiar and significant object and enabled astrophysicists to offer a clear and reasonable working model of the system. Yet important details are still unclear. For example, what is the mass of the compact star? This value would indicate whether this is a neutron star or if a black hole can be expected here. We also need more details for a better understanding of the acceleration of the jets and the collimation mechanisms. The next few years of observations will probably furnish additional details that may provide answers to these questions, or may even raise new questions for which we will need more answers.

12.6 Chaos in Stellar Variability

Chapter 5 treated stellar pulsations by linearizing the differential equations describing the structure and evolution of stars. From this method, which strictly maintains the limits on the validity of the linearized treatment, we obtained a basic understanding of the system. However a recently developed powerful mathematical technique makes possible the treatment of the nonlinear equations, thereby widening our ability to understand the variety of phenomena in nature.

Nonlinear dynamics and Chaos have, in the recent decade or so, become two of the most vigourously pursued research fields in physics. For a popular review we refer the reader to the best-selling book of Gleick.[24] Lately it has become clear that we can better understand various complex phenomena in nature by applying these new mathematical techniques. The striking discovery was that Chaos has an interdisciplinary nature and generic behaviour, and occurs in a wide range of scientific disciplines including mechanics, electrical circuits, lasers, fluid dynamics and turbulence, chemical reactions, biological systems, and even in economy and the social sciences.

Irregular time variability of various physical parameters (such as noise, many degrees of freedom, and complicated equations), previously thought to arise from very complex causes, may actually arise from rather simple physical systems with as little as three degrees of freedom. The basic cause for the seemingly complex behaviour was the inherent nonlinearity present in the physical system. Researchers realized that they can somtimes model complex behaviour using a low-dimensional fractal attractor (see Gleick).[24] An attractor, in the context of dynamic system theory, is a geometric structure reflecting the nature of the differential equations describing the dynamic system. We shall not go into the details of the abstract geometrical theory of differential equations. We shall try however to explain briefly the notion of phase space and attractors in this space with the help of an example.

As an example, let us consider a pendulum. This is a dynamic system which, by using Newton's second law, we can describe using a second order differential equation. The unknown variables are the displacement and velocity of the pendulum's bob as functions of time. It is possible to imagine a space spanned by the displacement and velocity (the phase space), and to view the motion of the pendulum as a trajectory in this space. It is well known that an ideal pendulum (like any undamped oscillator) describes an ellipse in phase space. The system "moves" along this ellipse as time passes. If friction is present, the trajectory is no longer a closed ellipse but spirals into a point (recalling a satellite in orbit, slowly losing energy by friction). Thus, irrespective of initial conditions, the trajectory in phase space approaches a point. In the case of the damped pendulum it is the equilibrium point (velocity and displacement from equilibrium are both equal to zero). Thus the stable equilibrium of the pendulum acts as an attracting point in phase space. It "attracts" all possible trajectories. In more complicated systems, the attractor can be a closed loop; or if involving more than one degree of freedom (a case in which the phase space is four-dimensional or more), attractors can be surfaces or volumes.

As pointed out above, many nonlinear dynamic systems — some of them rather simple — can exhibit a very complex behaviour, provided their phase space is three-dimensional or more. They are called Chaotic if the attractor exhibits a special and highly intriguing structure. It turns out that in these cases the attractor is a set of points with a fractal dimension. For example, in a Chaotic system with a three-dimensional phase space, the attractor will not appear as a point (a zero-dimensional set), nor as a closed curve (a one-dimensional set), or as a surface (a two-dimensional set). Its dimension is less than the phase space dimension (three). In such a case the number of the dimensions of the attractor is a noninteger number. We describe such structures in a branch of mathematics called *fractal geometry*. For a popular account of fractals, we refer the reader to Gleick's book.

Usually the differential equations describing a dynamic system contain one or more parameters reflecting the physical conditions. In the example of the pendulum, the length of the pendulum's string is such a parameter. In general, Chaotic behaviour appears only for some values of the parameters. In the case of one parameter, the system may behave regularly for a certain value of the parameter, but as the value of the parameter increases the behaviour of the system may become Chaotic. We have already stated that a damped pendulum is not Chaotic. Its attractor is a point. If we add a small force term to the equation, thereby extending the dimensions of the phase space to three, the trajectory is still attracted to a point. However if the magnitude of the force increases, the pendulum may start to move Chaotically and a fractal attractor appears. The qualitative change in the time behaviour appearing in this case due to the change in the attractor's dimensions is called a *bifurcation*.

The approach to Chaos which appears when changing a parameter is called a route to Chaos. One such famous route is the *Feigenbaum period-doubling bifurcations* discovered by M. Feigenbaum in 1978 and verified in convection experiments by A. Libchaber. In this scenario, when a parameter is varied, the system behaves periodically; but the system develops an oscillation with two frequencies, and later with four frequencies, eight frequencies, and so on, until Chaos appears. In the case of convection, the varied parameter (the "control" parameter) is the temperature difference in a layer of fluid heated from below. As the temperature difference increases, the regular rolls (which are periodic) develop secondary wiggles (two periods) and so on, until reaching a turbulent behaviour. In another route (found by Pomeau and Manneville) irregular bursts interupt the fundamental regular limited cycles when a parameter is varied, until the behaviour becomes fully Chaotic.

The models of pulsating stars (see Chapter 5, Section 5.6) offer a natural candidate for a dynamic system described by nonlinear differential equations, in which Chaotic behaviour may appear. Indeed if we give up the linearization of the equations for the perturbations and retain the nonlinear terms, we can expect complex time behaviour. Before the development of nonlinear analysis for dynamical systems, observers and theorists alike were biased in favour of the regularly pulsating stars, which could be well described by the usual linear analysis. Thus they have thoroughly studied the Cepheids, the RR Lyrae stars, and other periodic pulsators. However they also discovered and classified several types of nonperiodic pulsators as semiregular and irregular variables (e.g., W Virginis, Miras, and RV Tauri stars).

Irregular intrinsic stellar pulsation was the "step-child" of stellar pulsation theory. In fact the prevailing belief was originally that limited cycles, or at most double-mode pulsations, are the only attractors of the hydrodynamic equations. Only by the early 1980s did Buchler and Regev[25] and others propose the first simplistic models of stars exhibiting low-dimensional Chaotic behaviour. It is only in the last few years that they have undertaken a systematic study of the subject. We refer the reader to a recent review[26] for details and references. We shall summarize here only the main findings.

In models of certain pulsating stars, the effective temperature serves as the control parameter. When scientists lowered this temperature in models of intermediate luminosity stars, pulsations of increasing complexity arose, going through a period-doubling sequence (the Feigenbaum route) to Chaos. For high luminosity models, intermittent irregular bursts appeared until the pulsations became fully Chaotic (the Pomeau-Manneville route). Low luminosity models remained stable throughout. J.R. Buchler and coworkers created the models used in this work for the so called W Virginis stars, which are similar to regular Cepheids but have a chemical composition of population II stars. Classical Cepheids have a population I composition. It is remarkable that two of the universal routes to Chaos found in rather simple mathematical models (such as the Rossler and the Lorentz systems) should also appear in the complex-looking pulsating star models. The occurrence of Chaos proved to be a very robust feature of the models, and it develops a range of parameters of stellar models that put it in approximately the right place in the H-R diagram. It is very suggestive that the observed RV Tauri, Miras, and other semiregular variables owe their existence to an underlying low-dimensional fractal attractor.

Work on this subject is continuing and we expect more progress. The hope is that detailed analysis of data, using of new and sophisticated techniques, will soon shed new light on the behaviour of irregular stellar pulsators. On the theoretical side, scientists must systematically scan the H-R diagram by constructing detailed numerical stellar models and checking their behaviour, with an aim to reveal their Chaotic character if such is the case. In addition to these relatively "brute force" approaches, one may hope for the development of robust, relatively simple models which can extract the richness of the nonlinear behaviour of pulsating stars, thereby shedding light on the physical mechanisms responsible for their Chaotic variability.

References

1. Bahcall J.N., 1978, *Rev. Mod. Phys.*, **50**, 888.
2. Davis R., 1978, in *Proceeding of Informal Conference on Status and Future of Solar Neutrino Research*, ed. G. Freedlander, Brookhaven Nat. Lab. Report No. 50879, Vol. 1, Brookhaven.
3. Bahcall J.N., Ulrich R.K., 1988, *Rev. Mod. Phys.*, **60**, 297.
4. Davis R., 1988, Talk given at *Neutrino 88*, Tufts University.
5. Spergel D.N., Press W.H., 1985, *Astrophys. J.*, **294**, 663.
6. Finzi A., Harpaz A., 1989, *Astron. Astrophys.*, **211**, 441.
7. Barish B.C., 1990, in *Weak Interactions and Neutrinos*, ed. P. Singer, G. Eilam, North-Holland, Amsterdam.
8. Anselmann P., Hampel W., Heusser G., Kiko J., Kirsten T., Pernicka E., Plaga R., Ronn U., Sann M., Schlosser C., Wink R., Wojcik M., Ammon R.V., Ebert K.A.H., Fritsch T., Hellriegel K., Henrich E., Stieglitz L., Weyrich F., Balata M., Bellotti E., Ferrari N., Lalla H., Stolarczyk T., Cattadori C., Cremonesi O., Fiorini E., Pezzoni S., Zanotti L., Feilitzsch F.V., Mosbauer R., Schanda U., Berthomieu G., Schatzman E., Carmi I., Dostrovsky I., Bacci C., Belli P., Bernabei R., d'Angelo S., Paoluzi L., Charbit S., Cribier M., Dupont G., Gosset L., Rich J., Spiro M., Tao C., Vignaud D., Hahn R.L., Hartmann F.X., Rowley J.K., Stoenner R.W., Weneser J., 1992, *Physics Letters, B*, **285**, 376.
9. Shapiro S.L., Teukolsky S.A., 1983, *Black Holes, White Dwarfs, and Neutron Stars*, John Wiley & Sons, New York.
10. Regev O., 1975, M. Sc. thesis, Tel-Aviv University, Tel-Aviv.

11. Canuto V., 1974, *Ann. Rev. Astron. Astrophys.*, **12**, 167, and **13**, 335.
12. Baym G., Pethick C.P., 1979, *Ann. Rev. Astron. Astrophys.*, **17**, 415.
13. Hewish A., Bell S.J., Pilkington J.D.H., Scott P.F., Collins R.A., 1968, *Nature*, **217**, 709.
14. Taylor J.H., Stinebring D.R., 1986, *Ann. Rev. Astron. Astrophys.*, **24**, 285.
15. Rankin J.M., 1990, *Astrophys. J.*, **352**, 247.
16. Eichler D., Levinson A., 1988, *Astrophys. J. Lett.*, **335**, L67.
17. Johnston S., Manchester R.N., Lyne A.G., Bailes M., Kaspi V.M., Guojun Q., D'Amico N., 1992, *Astrophys. J. Lett.*, **387**, L37.
18. Clark D.H., 1985, *The Quest for SS 433*, Viking Penguin Inc., New York.
19. Milgrom M., 1979, *Astron. Astrophys.*, **76**, L3.
20. Fabian A.C., Rees M.J., 1979, *Mon. Not. Roy. Astron. Soc.*, **187**, 13.
21. Margon B., 1984, *Ann. Rev. Astron. Astrophys.*, **22**, 507.
22. Margon B., Anderson S.F., 1989, *Astrophys. J.*, **347**, 448.
23. Shapiro P.R., Milgrom M., 1986, *Astrophys. J. Supp.*, **60**, 393.
24. Gleick J., 1987, *Chaos*, Penguin Books, New York.
25. Buchler R.J., Regev O., 1982, *Astrophys. J.*, **263**, 312.
26. Buchler R.J., Regev O., 1990, in *The Ubiquity of Chaos*, AAAS, Washington.

Chapter 13

The Galaxy

Most of the stars we observe are part of the Galaxy. The name *Galaxy* originated from the Greek word for milk (*gala*). We observe the Galaxy in the night sky as a *Milky Way,* or a bright strip, which represents a high concentration of stars. The general structure of the Galaxy is of a nucleus and a disk embedded in a low density spherical-like structure (*spheroid*). The solar system exists in the disk, at about three-fifths of its radius. We observe the bright strip of the Milky Way when we look in the direction of the plane of the disk. Looking in any direction other than that of the disk plane, we see a much lower density of stars.

Our Galaxy is one of about 10^{11} galaxies observed in the part of the universe covered by our telescopes. Most of the galaxies are elliptical, and only few percent of them are similar to our Galaxy, which is a spiral galaxy. The elliptical galaxies contain mainly stars similar to population II (see Chapter 1), while spiral galaxies contain many stars of population I.

The following are the main features of the Galactic structure:

(1) The bulge, which contains the Galactic nucleus and its surroundings. This is an approximately spherical system with a radius of about 3,500 LY.

(2) The thin disk, which has a radius of about 50,000 LY and a height of around 2,500 LY.

(3) The thick disk, which has a height of about 8,000 LY and a lower star density than the thin disk.

(4) The spheroid, which is an oblate sphere with an average radius of about 40,000 LY.

(5) The halo, which is the matter surrounding the spheroid. It has a much lower density than the average density of the Galaxy, and its observational properties are vague.

13.1 The Early Phase[1]

Galaxies form from matter accumulated because of mutual gravitational attraction between its components. In diffused matter which is spread uniformly in space, the average gravitational interaction vanishes due to the isotropy of the system, and such a state can remain stable for long periods. However under certain conditions, local perturbations in the uniform distribution may grow to form nuclei for further gravitational collapse.

A perturbation of this kind may form a gravity centre. In a cloud of gas around such a centre, the particles bound to the centre are those for which the thermal energy, manifested in their dispersion velocities, is lower than the absolute value of their gravitational energy. The thermal energy per particle is of the order of kT, where k is Boltzmann constant and T is the average temperature of the gas cloud. The gravitational energy of a particle equals $\frac{-GMm}{r} = -G\frac{4\pi}{3}\rho r^2 m$, where r is the distance of the particle from the gravity centre and m is its mass. M is the mass enclosed in a sphere of radius r, and, in a distribution which is not far from uniform density, $M = \frac{4\pi}{3}\rho r^3$. Comparing the expressions for the thermal and gravitational energies, we find that in a matter cloud of (almost) uniform density, the maximal distance from which matter particles are still bound to the gravitational centre is given by:

$$r_{max} = \sqrt{\frac{3kT}{4\pi G\rho m}}.$$

If the mass collapses as a result of gravity without losing any energy, the gravitational energy released in the collapse converts to a kinetic energy of the particles which is expressed as an increase in the temperature of the gas. According to the virial theorem, the thermal energy of the new configuration is one-half of the absolute value of the gravitational energy, and the average radius of the new configuration is one-half of the original radius of the gas cloud. Designating the dispersion velocity of the particles by σ, we have: $m\sigma^2 \simeq kT$.

The temperature in the Galactic disk today is much lower than the virial temperature, which means that much energy was lost from the system. Energy can be lost from the system by radiation. This can happen when the matter heats up through inelastic collisons, and the atoms are raised to excited energy levels. When the atoms relax to their ground energy levels, the excess energy radiates away.

When most of the galaxies were formed, the average temperature of the matter in the universe was about 10^4 degrees, the hydrogen was fully ionized, and dispersion velocities were about 10 km sec^{-1}. During the collapse, gravitational energy converted to kinetic energy of the particles, which acquired supersonic velocities. Shock waves formed which brought about rapid heating and excitation of atomic energy levels. Loss of energy by radiation due to the atomic processes created the binding energy of the system. The time scale for such a process is the cooling time, t_{cool}, which characterizes the rate of energy radiation from the excited atoms.

For convenience we define the radiation cooling rate by $n^2\Lambda(T)$, where n is the number density of the particles, and $\Lambda(T)$ results from the characteristic processes which create the radiation. The processes of cooling depend on the type of atomic transitions involved — in other words whether these are free-free, free-bound, or bound-bound transitions (recall opacity as defined in Section 3.6). The last two types of transitions are more efficient in cooling than the free-free transitions. Bound-bound and free-bound transitions occur mainly in elements which are heavier than hydrogen and helium. Thus the percentage of heavy elements in the matter determines the cooling rate. The cooling time increases with the energy content of the cloud. The energy content is proportional to the density and the temperature of the cloud; however, the collision rate between particles increases with the second power of the density and results in a decrease in the cooling time. Hence the cooling time scale, t_{cool}, is given by:

$$t_{cool} = \frac{3nkT}{n^2\Lambda(T)} = \frac{3kT}{n\Lambda(T)}. \tag{13.1}$$

We should compare this time scale with the characteristic time for a free-fall in this system, t_{ff}. We can calculate t_{ff} if we recall that in a uniform density sphere, free-fall acceleration is proportional to the radius:

$$\ddot{r} = -\frac{4\pi}{3}G\rho r. \tag{13.2}$$

Integrating this equation twice yields:

$$r = R_0 \cos\frac{t}{t_{ff}} \tag{13.3}$$

where R_0 is the initial radius of the system, and $t_{ff} = (\frac{4\pi}{3}G\rho)^{-1/2}$. Equation 13.3 defines an oscillatory motion which would have taken place if no energy were lost from the system, and the density remained constant. Since energy is lost by radiation, the system is not conservative. The density increases with the contraction, the motion is not oscillatory, and t_{ff} is used only as an order of magnitude of the characteristic time of the collapse. This characteristic time, t_{ff}, depends on the density only. In a characteristic gas cloud which is supposed to form a galaxy, t_{ff}, is given by:

$$t_{ff} = 2 \times 10^7 n^{-1/2} \text{ yr.} \qquad (13.4)$$

A collapse is called "rapid" if its characteristic time is of the order of t_{ff}. Rapid collapse occurs when $t_{cool} < t_{ff}$.

In a temperature-density diagram we can plot a curve for which $t_{cool} = t_{ff}$. We find that for a temperature of 10^4 degrees, such a curve corresponds to a gas cloud with a mass of $10^{12} M_\odot$ and a radius of 300,000 LY, which are the upper limits for masses and radii in most of the observed galaxies. Most observed galaxies are located in the diagram below this curve, while clusters of galaxies lie above this curve. Individual calculations of these time scales for the Galactic bulge show that it, too, lies below this curve.

The scenario for the Galaxy's formation is that, starting with a free-fall, the cooling time was shorter than the free-fall time. The collapsing matter underwent sufficiently rapid cooling so as to allow for the dissipation and radiation of the excess energy from the system. A question therefore arises about how galaxies with higher masses were formed. The solution may be that these galaxies indeed maintained themselves for long periods in a quasi-steady-state while they cooled down slowly. This cooling process continued until increased density and decreased temperature reduced their cooling time scale and drove them into a rapid collapse. Jean's mass is defined as the upper limit for masses which are stable against density perturbations. In the case of density perturbations with masses above Jean's mass, gravity overcomes the pressure, and the mass collapses. This mass is given by:

$$M_J \simeq 10^2 T^{3/2} n^{-1/2} \ M_\odot. \qquad (13.5)$$

Masses of this order of magnitude are unstable against density perturbations, and they collapse. During the steady-state, Jean's mass was $10^6 M_\odot$, and this mass characterizes the mass range of the globular clusters observed today. Most of the globular clusters probably formed during this stage. In high mass galaxies, which existed for long periods in the steady-state, most of the stars formed before the galaxy collapsed.

The stars that make up the spheroid formed during the first stage of the rapid collapse of the proto-Galaxy. Matter that did not form stars and remained as gas underwent dissipation processes by which it rearranged itself in the form of a disk. The part of the matter which formed stars before the collapse underwent much less dissipation than the gaseous matter. This is because the ratio of surface to mass in stars is lower by few orders of magnitude than in gas. Hence the stars formed before the dissipation retained their original, nearly radial orbits and distributed isotropically to form the spheroid. Meanwhile the gaseous matter lost energy, and centrifugal force constrained it into the form of a disk.

In this picture, the time scale of the collapse corresponds to the time needed for a radial orbit, which is about 10^8 yr.

13.2 Chemical Evolution

The production rate for heavy elements through processes that take place in certain phases of stellar evolution can serve as a clock for recording the evolution.

As we know, supernovae have an important role in the production of heavy elements and their distribution in the interstellar medium. Elements such as oxygen, neon, and silicon form only in high mass stars which end their life as SN II. Low mass stars also produce elements of the iron group during SN I events. Recall from Chapter 10 that SN II events produce elements of the iron group and those in the range of the atomic mass of oxygen in a ratio of 1 : 2, whereas most of the heavy elements produced in SN I are of the iron group. An SN II explodes shortly after the formation of its progenitor star, while an SN I explodes about 10^9 yr after the formation of its progenitor star. Thus, at the first evolutionary phase, most of the heavy elements formed in SN II, and the ratio of iron to oxygen (Fe/O) was low. Only when significant number of SN I exploded did this ratio begin to increase. Hence the ratio Fe/O in existing stars gives an indication of stellar evolution in earlier epochs. Fe/O increases with time because the ratio of SN I events to SN II events increases with time.

SN I form from CO white dwarf stars located in binary systems. The stars ending their lives as CO white dwarfs were stars with masses in the range of 1 to 5 $M_\odot$ during their main sequence phase. The main sequence lifetime for such stars is in the range of 2.5×10^8 to 10^{10} yr. We therefore expect an increase in the Fe/O ratio after the first 10^9 yr of the Galaxy life. This means that population II stars, which contain a very low percentage of heavy elements, formed during the first phase of the life of the Galaxy, the phase of rapid collapse before the enrichment of the interstellar medium by heavy elements.

SN II occur in high mass stars, which have a very short lifetime on the order of 10^6 to 10^7 yr. Thus the rate of SN I events at a given epoch indicates the rate of stellar formation 10^9 to 10^{10} yr earlier; while the rate of SN II at a given epoch indicates the rate of star formation in that same given epoch. Comparing these rates, we conclude that the rate of star formation did not change much during the lifetime of the Galaxy and is about $10M_\odot$ yr^{-1}.

13.3 The Structure of the Galaxy

The spheroid is supposed to be the oldest component of the Galaxy and is composed of stars that formed during the phase of rapid collapse. We estimate the age of the spheroid as 15 Gyr (1 Gyr = 10^9 yr). Evidently the high mass stars of this generation completed their lifetime long ago, and the highest mass of stars of this population today is around $0.8M_\odot$. These are also low luminosity stars, and their spectral type is M and redder. They are designated as "extreme population II." In the vicinity of the solar system they are represented by subdwarf stars possessing high velocities. Taken together they form a spheroid, with the ratio between the polar and equatorial radii being about 0.5.

The Thin Disk

The best known part of the Galaxy is the regular (thin) disk. It has a height of about 2,500 LY, and most of the young luminous stars of the Galaxy exist within its dimensions. The disk formed from the dissipating gaseous constituent of the Galaxy as a result of the centrifugal force created by the Galaxy's rotation. The spheroid is "pressure supported" because it sustains its shape with the high isotropic radial velocities of its constituents. We regard the disk as "angular momentum supported" because it sustains its shape as a result of the conservation of the angular momentum of its constituents. The solar system lies at a radius of 28,000 LY, and the rotational velocity of the Galaxy at this location is about 200 km sec^{-1}. The solar system completes one circle in about 2.5×10^8 yr. The gas and the dust contained in the Galaxy reside in the disk. Here new stars are still being born today.

Pictures taken of disk galaxies other than the Milky Way show spiral arms in the disk. We also see the same features in our Galaxy when carrying out detailed star counting and mappings of the disk's three-dimensional structure. All the bright young stars lie in these arms, usually close to the gas and dust clouds from which they formed. It is clear that most of the activity in stellar formation takes place in the arms.

These arms are not solid structures. They are actually density waves that travel in the Galaxy. A density wave is a traveling front of increased density and pressure in which shock waves form and heating takes place. The increased density results in local condensations in the form of huge gas clouds. The merging of such clouds along their magnetic field lines creates supercritical masses which sets the stage for rapid collapse and results in the formation of high mass stars. The presence of bright high mass stars in the spiral arms makes the arms so pronounced, although the density excess in the arms is not much higher than the average density in the disk. The lifetime of the high mass stars is short, and they complete their life during the time they reside in the spiral arms. When the density wave proceeds, it leaves behind only low mass stars which constitute the field star population of the disk.

The density waves form from disturbances in the gravitational field of the Galaxy. A disturbance of few percent has the effect of an increase of tens of percent in the density. The disturbance waves in gravity result from the differential rotation of the Galaxy.

The Thick Disk

The proposal for the existence of the thick disk in the Galaxy occurred only recently.[1] Its height is about 8,000 LY (or about three times the height of the thin disk). In the region of the thin disk where both disks overlap, the stars of the thick disk make up a few percent of the common population of the two disks.

The population of the thick disk consists of a medium population II. The metallicity of this population is about a quarter that of the Sun. The vertical component of the velocity of these stars is about 45 km sec^{-1}, while that of the thin disk stars is about 22 km sec^{-1}. The rotation velocity of the thick disk stars is approximately 180 km sec^{-1}, and that of the spheroid stars is 40 to 50 km sec^{-1}. We should compare these velocities to the rotation velocity of the solar system, which is about 200 km sec^{-1}. Hence the thick disk is an "angular momentum supported" system. We estimate the age of the thick disk population to be 8 to 11 Gyr.

The thick disk probably formed as the initial state before the formation of the thin disk. After the rapid collapse during which the spheroid stars formed, the important process became the dissipation of the thermal energy. During this process the gaseous matter began to form the Galactic disk as a result of the conservation of angular momentum. At this stage the cooling rate increased significantly because the supernovae explosions of stars of the first generation enriched the interstellar medium with heavy elements. The consequent higher metallicity resulted in a higher cooling rate. This acceleration of cooling caused star formation during the transition from a spheroid to a disk structure. The stars formed during this transition lie in the thick disk.

The Bulge

The bulge is an almost spherical structure around the nucleus of the Galaxy.[2] Its radius is about 3,500 LY, and it has a mass of about $10^{10} M_\odot$. The average matter density of the bulge is a hundred times the average density of the Galaxy. This structure superimposes on the structures of the spheroid and the two disks.

We find the age of a stellar population from the age of the brightest stars on the main sequence of the population. The masses of the brightest main sequence stars in the bulge population are below $1 M_\odot$. Their age is approximately 11 Gyr, which is similar to the age of the thick disk stars. Surprisingly, the metallicity of the bulge stars is twice that of the Sun and about 10 times that of the thick disk stars. The explanation for this is that due to the much higher matter density in this region, the rate of star formation is high as well, with probably a corresponding higher fraction of high mass stars in the population. This situation results in a higher ratio of supernova events which enrich the interstellar medium with heavy elements. Thus when the present bulge stars formed 11 Gyr ago, the metallicity in the interstellar medium at their location was twice that of the solar vicinity when the Sun formed 4.6 Gyr ago.

The Nucleus

It is at the centre of the bulge that the nucleus of the Galaxy resides. Due to the presence of thick dust and gas in the Galactic plane, the central part cannot be observed in the range of optical, UV, or soft X-ray wavelengths. The range of wavelengths that can penetrate the dust and gas in the disk plane are radio, infrared, and γ rays. Scientists observed radiation of γ photons with energies above 0.51 MeV. This energy equals the rest energy of the electron. The presence of these photons shows that processes of creation and annihilation of electron-positron pairs take place there as well.

Figure 13.1 displays a schematic model of the circumnuclear disk. The main observations are in the range of radio radiation. These observations, from which contour maps were drawn, progressed significantly in the last

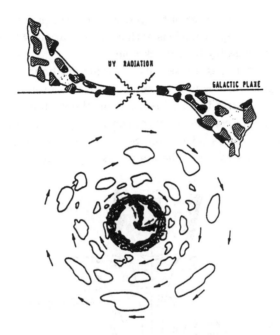

Figure 13.1. Schematic model of the circumnuclear disk. [Adopted from Genzel.[4]]

decade. A ring of ionized hydrogen with a radius of about five LY lies[3] at the centre of this region. The ring may be the product of a spherical structure that flattened due to rotation, and whose centre was cleared of dust and gas by powerful radiation. The ring seems to exist in a steady-state. The matter around the ring contains clouds of molecular hydrogen in which old stars exist. However, young, bright stars are also observed. This means that there the process of star formation is still taking place. Researchers also observed several X-ray sources which are probably close binary systems. Another interesting observation is the existence of elongated structures that perpendicularly align to the Galactic plane and stretch to a length of few hundred LY. This observation suggests the presence of a magnetic field at the Galactic centre. It is not yet clear how much this field influences the processes occurring in this region.

At the very centre of the ring, scientists note a strong compact radio source having a diameter of less than one light hour. This source is Sagitarius A* and is considered to be the nucleus of the Galaxy. Analysis of the peculiar velocities of the nucleus' surrounding clouds shows that its gravity corresponds to a mass of the order of few times 10^6 $M_\odot$. The general picture is of a compact nucleus with a mass of about 4×10^6 $M_\odot$. This central mass may possibly be a black hole.

Around this system lies a disk with a radius of about 24 LY. The disk contains hydrogen which is mostly recombined. It is probably an accretion disk formed from matter flowing inward from the vicinity.

To demonstrate the complexity of the central region we present in fig. 13.2 a sketch of the region, as taken from Sofue.[5] The distances in this sketch are measured in parsecs (pc), where a parsec equals about 3.1 LY.

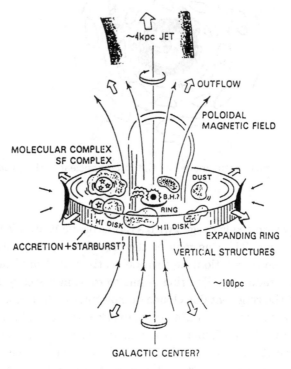

Figure 13.2. Schematic view of the "Galactic centre." [Adopted from Sofue.[5]]

We see in the figure the location of the proposed black hole at the centre with the ionized ring surrounding it. The long vertical arrows represent the magnetic field, and the accretion disk encircles the whole system at a radius of 24 LY. Molecular clouds which give birth to new stars reside both inside and outside the accretion disk.

References

1. Gilmore G., Wyse R.F.G., Kuijken K., 1989, *Ann. Rev. Astron. Astrophys.*, **27**, 555.
2. Frogel J.A., 1988, *Ann. Rev. Astron. Astrophys.*, **26**, 51.
3. Townes C.H., 1989, in *The Center of the Galaxy*, ed. M. Morris, Kluwer Academic Press, Dordrecht.
4. Genzel R., 1989, in *The Center of the Galaxy*, ed. M. Morris, Kluwer Academic Press, Dordrecht.
5. Sofue G.Y., 1989, in *The Center of the Galaxy*, ed. M. Morris, Kluwer Academic Press, Dordrecht.

Index